Strategic Applications of Measurement Technologies and Instrumentation

Soubantika Palchoudhury
University of Tennessee at Chattanooga, USA

A volume in the Advances in Chemical and Materials Engineering (ACME) Book Series

IGI Global
DISSEMINATOR OF KNOWLEDGE

Published in the United States of America by
IGI Global
Engineering Science Reference (an imprint of IGI Global)
701 E. Chocolate Avenue
Hershey PA, USA 17033
Tel: 717-533-8845
Fax: 717-533-8661
E-mail: cust@igi-global.com
Web site: http://www.igi-global.com

Library of Congress Cataloging-in-Publication Data

Names: Palchoudhury, Soubantika, 1985- editor.
Title: Strategic applications of measurement technologies and instrumentation / Soubantika Palchoudhury, editor.
Description: Hershey, PA : Engineering Science Reference, [2018] | Includes bibliographical references and index.
Identifiers: LCCN 2017043496| ISBN 9781522554066 (h/c) | ISBN 9781522554073 (eISBN)
Subjects: LCSH: Measuring instruments. | Measurement.
Classification: LCC QC100.5 .S735 2018 | DDC 530.8--dc23 LC record available at https://lccn.loc.gov/2017043496

This book is published in the IGI Global book series Advances in Chemical and Materials Engineering (ACME) (ISSN: 2327-5448; eISSN: 2327-5456)

British Cataloguing in Publication Data
A Cataloguing in Publication record for this book is available from the British Library.

Advances in Chemical and Materials Engineering (ACME) Book Series

ISSN:2327-5448
EISSN:2327-5456

Editor-in-Chief: J. Paulo Davim, University of Aveiro, Portugal

MISSION

The cross disciplinary approach of chemical and materials engineering is rapidly growing as it applies to the study of educational, scientific and industrial research activities by solving complex chemical problems using computational techniques and statistical methods.

The **Advances in Chemical and Materials Engineering (ACME) Book Series** provides research on the recent advances throughout computational and statistical methods of analysis and modeling. This series brings together collaboration between chemists, engineers, statisticians, and computer scientists and offers a wealth of knowledge and useful tools to academics, practitioners, and professionals through high quality publications.

COVERAGE

- Polymers Engineering
- Industrial chemistry
- Artificial Intelligence Methods
- Nanomaterials
- Sustainable Materials
- Ductility and Crack-Resistance
- Numerical Techniques
- Metallic Alloys
- Electrochemical and Corrosion
- Statistical methods

IGI Global is currently accepting manuscripts for publication within this series. To submit a proposal for a volume in this series, please contact our Acquisition Editors at Acquisitions@igi-global.com or visit: http://www.igi-global.com/publish/.

The Advances in Chemical and Materials Engineering (ACME) Book Series (ISSN 2327-5448) is published by IGI Global, 701 E. Chocolate Avenue, Hershey, PA 17033-1240, USA, www.igi-global.com. This series is composed of titles available for purchase individually; each title is edited to be contextually exclusive from any other title within the series. For pricing and ordering information please visit http://www.igi-global.com/book-series/advances-chemical-materials-engineering/73687. Postmaster: Send all address changes to above address.

Titles in this Series

For a list of additional titles in this series, please visit:
https://www.igi-global.com/book-series/advances-chemical-materials-engineering/73687

The Geometry of Higher-Dimensional Polytopes
Gennadiy Vladimirovich Zhizhin (Russian Academy of Sciences, Russia)
Engineering Science Reference • ©2019 • 286pp • H/C (ISBN: 9781522569688) • US $195.00

Composites and Advanced Materials for Industrial Applications
K. Kumar (Birla Institute of Technology, India) and J. Paulo Davim (University of Aveiro, Portugal)
Engineering Science Reference • ©2018 • 402pp • H/C (ISBN: 9781522552161) • US $225.00

Emerging Synthesis Techniques for Luminescent Materials
Ratnesh Tiwari (Bhilai Institute of Technology, India) Vikas Dubey (Bhilai Institute of Technology, India) and Sanjay J. Dhoble (Rashtrasant Tukadoji Maharaj Nagpur University, India)
Engineering Science Reference • ©2018 • 505pp • H/C (ISBN: 9781522551706) • US $215.00

Handbook of Research on Ergonomics and Product Design
Juan Luis Hernández Arellano (Autonomous University of Ciudad Juárez, Mexico) Aide Aracely Maldonado Macías (Autonomous University of Ciudad Juárez, Mexico) Juan Alberto Castillo Martínez (University of Rosario, Colombia) and Porfirio Peinado Coronado (Autonomous University of Ciudad Juárez, Mexico)
Engineering Science Reference • ©2018 • 446pp • H/C (ISBN: 9781522552345) • US $235.00

Mechanical Properties of Natural Fiber Reinforced Polymers Emerging Research ...
Sarah S. Gebai (International University of Beirut, Lebanon) Ali M. Hallal (International University of Beirut, Lebanon) and Mohammad S. Hammoud (International University of Beirut, Lebanon)
Engineering Science Reference • ©2018 • 228pp • H/C (ISBN: 9781522548379) • US $165.00

For an entire list of titles in this series, please visit:
https://www.igi-global.com/book-series/advances-chemical-materials-engineering/73687

IGI Global
DISSEMINATOR OF KNOWLEDGE

701 East Chocolate Avenue, Hershey, PA 17033, USA
Tel: 717-533-8845 x100 • Fax: 717-533-8661
E-Mail: cust@igi-global.com • www.igi-global.com

Table of Contents

Detailed Table of Contents

The effects of Salako nutritional supplements on cocaine-dependent Sprague Dawley rats was investigated. Rats were made cocaine-dependent using conditioned place preference (CPP) where craving was analyzed regularly. Cocaine metabolite levels were determined from blood samples. CYP450 3A4 isoenzyme activities were obtained using liver homogenate. Statistical analysis was done using SPSS one-way ANOVA and Duncans multiple range test. Results show that when cocaine use was discontinued, the supplements reduced craving of cocaine significantly. Blood plasma results showed higher benzoylecgonine equilibrium possibly indicating that the supplements aided the removal of stored cocaine metabolites which may have contributed to better management of craving in the rats. CYP450 3A4 isoenzyme activity was further enhanced by the supplements and is indicative of increased cocaine metabolism. The results indicate that the Salako nutritional supplements reduce craving caused by chronic cocaine administration by increasing the liver CYP450 3A4 isoenzyme activity, resulting in better plasma clearance.

The authors prove that the standard least action principle implies a more general form of the same principle by which they can state generalized motion equation including the classical Euler equation as a particular case. This form is based on an observation regarding the last action principle about the limit case in the classical approach using symmetry violations. Furthermore, the well-known first integrals of the classical Euler equations become only approximate first integrals. The authors also prove a generalization of the fundamental lemma of the calculus of variation and they consider the application in electromagnetism. This chapter is an enhanced version of a published work. It proves the existence of particular relativistic error condition in classical mechanics, potentially significant on experiments of light propagation in matters. The work includes a discussion of applications potentially correlated with the found particle motion error condition.

In this chapter, the wavelength dependence of bend loss in a step-index multimode optical fiber (100 μm core diameter; fused silica) was investigated for fiber bend radii ranging between 2.0 and 4.5 mm using six laser excitation wavelengths, namely, 337.1, 470, 590, 632.8, 750, and 810 nm. The results obtained from fitting the bend loss measurements to Kao's model and utilizing MATLAB® indicate that bend loss is wavelength dependent and transmission loss in multimode optical fibers increases with the decrease in the fiber bend radius. Furthermore, the response of a microbend fiber-optic displacement sensor was characterized at 337.1, 470, 632.8, 750, and 810 nm. Measurements obtained from the microbend sensor indicate that the sensor output power is linear with the applied displacement and the sensor output is wavelength dependent. Lastly, references for industrial and biomedical applications of microbend fiber-optic sensors are provided. Finally, a brief description for the transmission loss mechanisms in optical fibers is given.

Section 2
Measurement Strategies in Industrial Applications and Development of Suitable Reference Materials

This section describes the significance of measurement technology in industry through three model examples. Consequently, the need for appropriate reference materials and measurement methods is highlighted in the concluding chapters of this section.

This chapter deals with the diagnosis of induction motors (IM) with the so-called motor current signature analysis (MCSA). The MCSA is one of the most efficient techniques for the detection and the localization of electrical and mechanical failures, in which faults become apparent by harmonic components around the supply frequency. This chapter presents a summary of the most frequent faults and its consequences on the stator current spectrum of an IM. A three-phase IM model was used for simulation taking into account in one hand the normal healthy operation and in the other hand the broken rotor bars, the shorted turns in the stator windings, the voltage unbalance between phases of supply, and the abnormal behavior of load. The MCSA is used by many authors in literature for faults detection of IM. The major contribution of this work is to prove the efficiency of this diagnosis methodology to detect different faults simultaneously, in normal and abnormal functional conditions. The results illustrate good agreement between both simulated and experimental results.

The performance and maintenance of testing laboratories is a prime issue. The quality of coal test results not only depends on performance of individual results but it also depends on the performance of various tests in the same laboratory. Machine interference is a significant problem in many manufacturing system and testing equipment. The variation of results for testing equipment may be due to various factors which need to calculate the uncertainty of measurement to show the accuracy of the machine. In case of coal testing laboratory, the plant layout and surrounding environment affects the performance of the system. The machine interference comes under variable causes which may affect the result. This chapter proposes a methodology for constructing system performance measures, finding out the various factors responsible for variations in result. The chapter deals with estimation of machine interference existence using variable control chart approach for coal testing equipment. The analysis of results for such machine interference will be useful and significant for system designers and practitioners.

This chapter mainly focused on the recent trends in the antenna design techniques for next-generation wireless communication systems. Ultra-wideband antenna and multi-input-multi-output antennas are very useful to achieve higher data rates. An antenna is a transducer that changes guided electromagnetic energy in a transmission line to radiated electromagnetic energy in free space. Antennas may also be observed as an impedance transformer, coupling among an input or line impedance, and the impedance of free space. The looming widespread commercial deployment of ultra-wideband (UWB) systems has flashed new interest in the subject of ultra-wideband antennas. The power levels approved by the FCC mean that every dB counts in a UWB system as much as or possibly even more so than in a standard narrowband system. Thus, in effect, UWB antenna is a precarious part of an overall UWB system design. Basic principles for ultra-wide-band (UWB) antenna design and radiation are presented and discussed in this chapter.

Preface

Measurement techniques form the basis of scientific, engineering, and industrial innovations. The methods of measurement, standard or control for measurement, and instruments for measurement in different fields are constantly improving. A few such cases are presented here. This book describes the significant role played by measurement techniques in medical and scientific discoveries as well as modern industrial applications. It is a compilation of enhanced version of select articles published in the International Journal of Measurement Technologies and Instrumentation Engineering by IGI Global publishers. The book is divided into two sections, Significance of Measurement Strategies in Physics and Biomedical Applications (Section 1) and Measurement Strategies in Industrial Applications and Development of Suitable Reference Materials (Section 2).

Section 1 contains three chapters related to basic sciences, the first addressing the use of modern measurement techniques in healthcare applications, the second predicting a generalized variation of the most fundamental equation in physics, the principle of least action through mathematical derivation and measurement of errors and the third investigating the effect of wavelength on bend loss in step-index multimode optical fibers. Chapter 1 describes conditioned place preference in animal models as a measurement method to assess treatment strategies for cocaine addiction. Cocaine addiction is a critical issue as it can induce severe health problems in the patient like drastic weight loss, malnutrition and lesions. The key to address this issue is timely treatment and preventing a relapse of the addiction. Hospitals and medical centers across the globe provide programs including around-the-clock inpatient care, outpatient care or a mix of the two to treat cocaine addiction. However, current mode of treatment mostly uses behavioural intervention or learning strategies such as cognitive behavioural therapy to ensure abstinence of the patient from cocaine. A most suitable approach will be to integrate behavioural therapy with pharmacological treatment such as nutritional supplements. In chapter 1, conditional place preference in rat models is used as a measurement tool to prove the effectiveness of a nutritional supplement formulation in reducing cocaine craving. The finding is significant for the field of healthcare and medicine. Chapter

2 presents the use of a fundamental lemma of the calculus of variations to derive a generalized form of Euler equations. The classical Euler equation is shown as a special case of this generalized form. We know that the principle of least action is one of the most elegant equations in physics as it connects classical and quantum mechanics and can be used to derive any physical equation. The proof in Chapter 2 is significant in the field of physics because it shows that the standard form of the principle of least action implies a generalized variational principle which leads the derivation to the generalized Euler equation. Chapter 3 reports a fundamental finding related to bend loss in optical fiber through a detailed study of silica fused fiber under six different laser excitation wavelengths. The measurements obtained from microbend sensor reveal the wavelength dependence of bend loss in optical fibers. In Chapter 3, detailed analysis of bend loss using Kao's model fit and transmission loss mechanisms in optical fibers are also described. The work will be highly beneficial in further enhancing the micro-bend fiber optic sensor for more sensitive detection of displacements and cost-effective industrial applications.

Section 2 is dedicated to measurement and assessment methods used in industrial applications and the significance of developing certified reference materials. This section contains five chapters. Chapter 4 discusses the common mechanical and electrical faults found in induction motors and reports a conditioned monitoring and diagnostic approach using Motor Current Signal Analysis (MCSA) to prevent major failures of induction motors. Induction motor is one of the fundamental machines used in almost all industries. Therefore, study of suitable measurement methods to diagnose and prevent failure of induction motors is of high significance for industrial applications. In this study, a motor ring experimental set-up is devised to simulate four specific cases of motor fault including broken rotor bars, broken rotor bars and unbalanced supply voltage, broken rotor bars along with unbalanced supply voltage and load torque fluctuation and asymmetric stator winding. In each case the faults are analysed and detected using MCSA technique. In testing industrial equipment and materials, uncertainty of measurement is an important parameter. While computing the results from these tests follows specific and known steps, uncertainty evaluation is influenced by different factors and is more complex. Chapter 5 introduces a tool known as Uncertainty Budget for making realistic uncertainty measurements through a set of internationally recognized outlines. Chapter 6 focuses on the emerging trends in antenna design to meet the demand for higher data rate in wireless communication systems. In this chapter, a metric to define antenna design is established and emphasis is given on ultra-wide band and multi-input-multi-output antenna types. Therefore, this chapter covers the measurement strategies and techniques for developing next-generation wireless communication. Chapter 7 highlights the major role of certified reference materials in metrology. The chapter introduces the concept of reference materials as defined by the International Organization for Standardization (ISO).

Certified reference materials (CRM) are metrologically characterized standards from internationally recognized manufacturers. They provide a reliable control material for measurements in industrial testing and research. In addition, CRMs facilitate the report of comparable results from different laboratories across the globe. Therefore, CRMs should be of high quality. Chapter 7 presents different environmental CRMs developed at the National Metrology Institute from Brazil, Inmetro. In Chapter 8, the important issue of greenhouse gases is addressed from a measurement perspective. This chapter describes the need to develop better quality and monitoring points for greenhouse gases and the recent advances in the field. The methods described will be beneficial in better predicting climate and sources for the greenhouse gases.

This book compiles the cutting-edge research from scientific groups all over the world. Professors, researchers and industry professionals from different countries (e.g., United States of America, Jamaica, Brazil and India) have collaborated in this project. Therefore, the book will be highly useful for a global audience. In particular, graduate students and researchers will benefit from the new measurement techniques and modern applications in addiction treatment, specific mathematical solution to uncertainties in classical mechanics, industrial measurements and certified reference material described in the book. It will be most useful as a reference textbook for advanced courses in measurement and instrumentation. Target audiences will also include researchers and scholars in the field of healthcare, physics, industrial instrumentation, engineering and environmental science.

As the reader will appreciate, the wide range of modern topics covered in this book could not have been realized without the contribution of the expert authors. Accordingly, I would like to specially thank the authors and recognize their valuable contribution that made this publication possible.

Soubantika Palchoudhury
University of Tennessee at Chattanooga, USA

Acknowledgment

I would like to acknowledge the help of all the people involved in this project and, more specifically, the authors. Without their support, this book would not have become a reality.

First, I would like to thank each one of the authors for their contributions. Our sincere gratitude goes to the chapter's authors who contributed their time and expertise to this book.

Second, I would like to acknowledge the College of Engineering & Computer Science, University of Tennessee Chattanooga for continued support during the compilation of this book.

Soubantika Palchoudhury
University of Tennessee at Chattanooga, USA

Section 1

Significance of Measurement Strategies in Physics and Biomedical Applications

This section highlights two new measuring techniques used in the fields of human health and physics. The importance of modern measurement strategies in basic sciences is presented through three model cases in this section.

Chapter 1
Plasma Cocaine Metabolite Levels and Liver CYP450 3A4 Isoenzyme Activity as Indicators of Cocaine Metabolism in Rats Treated With Salako Supplements

Natwaine Sherune Gardner
University of the West Indies, Jamaica

Kedon J. S. Luke
University of the West Indies, Jamaica

Andrew O. Wheatley
University of the West Indies, Jamaica

Winston De La Haye
University of the West Indies, Jamaica

Perceval Steven Bahado-Singh
University of Maryland Medical System, USA

Lowell L. Dilworth
University of the West Indies, Jamaica

Donovan A. McGrowder
University of the West Indies, Jamaica

Everard Barton
University of the West Indies, Jamaica

Lauriann E. Young-Martin
University of the West Indies, Jamaica

Ajibeke Salako-Akande
Getwele Natureceuticals, USA

Henry Lowe
Environmental Health Foundation, Jamaica

Errol Morrison
National Commission on Science and Technology, Jamaica

Denise Eldermire-Shearer
University of the West Indies, Jamaica

Helen Asemota
University of the West Indies, Jamaica

DOI: 10.4018/978-1-5225-5406-6.ch001

ABSTRACT

The effects of Salako nutritional supplements on cocaine-dependent Sprague Dawley rats was investigated. Rats were made cocaine-dependent using conditioned place preference (CPP) where craving was analyzed regularly. Cocaine metabolite levels were determined from blood samples. CYP450 3A4 isoenzyme activities were obtained using liver homogenate. Statistical analysis was done using SPSS one-way ANOVA and Duncans multiple range test. Results show that when cocaine use was discontinued, the supplements reduced craving of cocaine significantly. Blood plasma results showed higher benzoylecgonine equilibrium possibly indicating that the supplements aided the removal of stored cocaine metabolites which may have contributed to better management of craving in the rats. CYP450 3A4 isoenzyme activity was further enhanced by the supplements and is indicative of increased cocaine metabolism. The results indicate that the Salako nutritional supplements reduce craving caused by chronic cocaine administration by increasing the liver CYP450 3A4 isoenzyme activity, resulting in better plasma clearance.

INTRODUCTION

Cocaine ($C_{17}H_{21}NO_4$), classified as a stimulant, is described as being the most potent, powerfully addictive stimulant of natural origin. The leaves of the Coca plant (*Erythroxylum coca*) can be harvested several times a year. The alkaloid, cocaine, is extracted from the leaves of the Coca plant, which originates in South America, and to a lesser extent, in Africa, Indonesia and India (UNODC, 2010). Cocaine is one of the oldest known psychoactive substances. Coca use has been traced as far back as around 5000 B.C. wherein the leaves of the plant were continually chewed in the mouth. Pure cocaine was isolated in the 1880's (National Institute on Drug Abuse, 2008). In the early 1900s, pure cocaine was the main active ingredient in numerous pharmaceutical and recreational formulations due to their properties that enhanced general activity and decreased fatigue.

Illicit cocaine from South America is produced normally as relatively pure hydrochloride salt (ranging from 80 – 95 percent) for export to the United States. Illicit cocaine is distributed ranging from a white crystalline powder (cocaine hydrochloride), to that of an off-white chunky material ("crack" or "rock" cocaine) (Drug Enforcement Agency, 2005). Street cocaine is usually adulterated with various substances such as mannitol, lactose and glucose, talc and flour. Due to the heat labile property of cocaine hydrochloride, the preferred method of introduction is usually via snorting or by dissolving in water and injected. Crack cocaine on the other hand is smoked (Drug Enforcement Agency 2005). Crack is the term given to the product formed with the processing of cocaine with ammonia or sodium bicarbonate and water.

Cocaine use varies widely with drug users and can range from occasional to repeated or compulsive use. The primary routes of cocaine administration are intranasal, intravenous (injection), inhalation and oral. Injecting the drug releases the drug directly into the bloodstream which increases the drug intensity. The duration of the euphoric effects produced by cocaine is dependent on the route of administration. The faster the absorption of cocaine, the more rapid and intense the high produced, but the shorter the duration of action. Toxic amounts of cocaine can be absorbed by any route of administration, which can have a wide range of effects which can include sudden death (Williams & Wasserberger 2006).

Metabolism of Cocaine

Cocaine is rapidly absorbed in the blood following the snorting, smoking and intravenous administration. Cocaine is metabolized by several enzymatic and non-enzymatic pathways (Heard *et al.*, 2008) rapidly and extensively in animals to yield benzoylecgonine and ecgonine as the principal products, along with other metabolites, particularly ecgonine methyl ester, which are all generally inactive. Norcocaine is a minor metabolite that is formed, but is active and found to be neurotoxic. Cocaine is converted to norcocaine via the cytochrome P450 3A4 isoenzyme by means of N-demethylation. Norcocaine has shown to be equal in potency to cocaine (Hawks *et al.*, 1974; Misra *et al.*, 1975).

The mechanism by which cocaine causes its central nervous system effects are more involved than that of the effects on the neurotransmitter system. The pharmacology of cocaine is complex, which exhibits simultaneous effects in several organ systems in the body including the brain, cardiovascular, immune and hematological systems (Espinoza, 2012; Heard *et al.*, 2008 and Levine *et al.*, 1990). Cocaine has been shown to be extremely hepatotoxic (Boyer & Peterson, 2005; Kanel *et al.*, 1989; Pasanen *et al.*, 1995) and also results in fatty infiltration responsible for increases in lipid content of the liver (Affi *et al.*, 1998; MachLachlan & Hodge, 1938). Chronic use of cocaine results in the metabolite being present in the urine for a prolonged period following introduction of a large amount into the body.

Cocaine was one of the first of the "natural" organic compounds to be chemically investigated. There is still a paucity of information which exists regarding the effects that cocaine has on the body as a complete system, including the toxicity that exists during chronic usage and the general basic mode of action following varying modes of administration. Developing an effective, safe and cost-effective means of treating substance abuse is of paramount importance in society.

Effects of Cocaine

Cocaine is a strong central nervous system (CNS) stimulant which performs its action by interfering with catecholamine reabsorption, especially that of dopamine (a chemical messenger that is associated with pleasure and movement), by blocking the dopamine transporter (Rebec, 2012). This then results in increased levels of extracellular dopamine, resulting in an increased level of dopamine being present in the synaptic gaps, thereby leading to a feeling of "rush", described as euphoria (Couper & Logan 2004). Cocaine is also involved in the blocking of norepinephrine and serotonin in the synaptic gaps. Following even a single dose, the effects of cocaine use can be seen almost immediately. In small dosages, cocaine gives the user a feeling of euphoria, making them feel talkative, energetic and mentally alert (especially to touch, sight and sound), and decrease the need for food and sleep (National Institute on Drug Abuse, 2008).

Short Term Effects

Cocaine affects the physiology short term by causing constricted blood vessels, dilated pupils, and increasing heart rate, body temperature and blood pressure. The user's high is intensified by increasing the amount of cocaine, but may sometimes lead to haphazard and violent behaviour. Users may exhibit a wide range of experiences including restlessness, paranoia, anxiety, panic, tremors, vertigo and muscle twitches (Tsai & Zager 2005). Medical effects include more commonly cardiovascular effects (such as abnormal heart rhythm and heart attack), neurological effects (such as strokes, seizures and coma) and complications with gastrointestinal system. Even when cocaine is used for the first time, risks can include sudden death which is often related to cardiac arrest of seizures followed by respiratory arrest (Bates, 1988).

Long Term Effects

The addictive rate associated with cocaine is high due to its powerfully reinforcing properties and the user is unable to control the extent to which the drug is craved (Robinson & Berridge, 2005). With repeated cocaine exposure, the brain adapts which makes the reward pathway become less sensitive, developing tolerance which requires increasing doses and/or more frequent use of cocaine in order to achieve the same pleasurable effect as in the initial use (Koob & Le Moal, 2001 & Gutman, 2006). This effect of tolerance leads to 'binges' where the drug user uses cocaine repeatedly and at increasingly higher dosages. Chronic users of cocaine experience significant weight loss and malnourishment (National Institute on Drug Abuse, 2008).

Treating Cocaine Addiction and Nutritional Supplements Administration

Cocaine addiction is described as being a complex disease, which involves biological changes to and in the brain, along with numerous other problems. The approach to treating cocaine dependence is a comprehensive method where the strategies employed assess the social, medicinal and neurobiological aspects of the patient on an individual basis. Presently there are no pharmacological methods of treating cocaine misuse (Garcia-Rodriguez *et al.*, 2009; Kennedy *et al.*, 2012). The medications that are used currently treat symptomatically the effects produced as a consequence of the discontinuation of cocaine use. There are numerous research areas that are investigating possible treatment methods for cocaine misuse however the key areas that show promise operate on the premise of the cocaine-induced adaptations that occur which disturb the balances between excitatory and inhibitory neurotransmission which primarily involve the glutamate and gamma-amino-butyric acid (GABA) (Hu, 2007; Volkow, 2009).

The main modality of the treatment of cocaine misuse involves behavioural interventions, the most effective being Cognitive-behavioral Therapy (CBT) (Garcia-Rodriguez *et al.*, 2009; Maude-Griffin *et al.*, 1998; Weiss R. D. *et al.*, 2005). CBT is focused on the abstinence of the user from cocaine and is based on the assumption that the development and continuation of cocaine misuse are influenced by learning processes (Center for Substance Abuse Treatment, 1999). The integration of both behavioural and pharmacological treatments will seemingly make the most effective approach. However, at present there is no medication that treats cocaine misuse (Kennedy *et al.*, 2012). There is therefore the need for effective pharmacological method of treatment of cocaine misuse.

The Salako Nutritional Supplements formulation used in this study (Provisional Patent# 60/528,755) was developed in the clinics of the Getwele Treatment and Rehabilitation Services Incorporated in Maryland, USA by Dr. Ajibike Salako-Akande for use in reducing craving for alcohol and illicit drugs. Cocaine treatment, with this proprietary formulation, involves the use of multi-nutritional formulation fortified with vitamins and minerals, which has been shown to be effective in relapse management and prevention. It is being screened in our laboratories to check if it may represent a safe and cost effective secondary prevention approach to cocaine addiction.

Conditioned Place Preference

Conditioned place preference is one popular method of an animal model of drug dependence. The apparatus used to carry out the experiment is called a conditioned place preference box, which consists of two chambers separated by a neutral chamber. One chamber differs from the other by tactile clues such as colour and light intensity. Using a biased procedure, the non-preferred compartment is paired with the drug and the CPP is a measurement of overcoming the initial aversion for that compartment (Sanchis-Segura & Spanagel, 2006). In adult rats, place preference for cocaine is effectively established for cocaine at a dosage of 20 mg/kg (Bardo et al. 1986; and Brenhouse & Andersen, 2008).

MATERIALS AND METHODS

Cocaine hydrochloride was obtained from Health Cooperation Limited, Kingston, Jamaica with permission from the Ministry of Health of Jamaica.

Kits for Cocaine hydrochloride determination were purchased from Neogen Cooperation Ltd., USA.

Kits for Cytochrome P450-3A4 isoenzyme determination were purchased from Promega Corporation, USA.

Salako Nutritional Supplements formulation was provided by Dr. Ajibike Salako-Akande of Getwele Foundation International, USA.

Animal Study

Design of Feeding Experiment

Cocaine hydrochloride acquisition and usage were with the approval of the Ministry of Health Dangerous Drugs Inspector, Jamaica. The animal study was performed with the approval of the University of the West Indies, Mona Ethics Committee. Fifty (50) adult Sprague-Dawley rats (25 males and 25 females) between the ages of 6-8 weeks were obtained from the University of the West Indies Animal House.

The rats were separated, weighed, labeled for differentiation; and placed into five groups (shown in Table 1), with an average body weight of 238.5 g. The stainless-steel cages that they were housed in were maintained daily and were placed in a room that was kept on a 12/12-hour light/dark cycle. Throughout the entire study, the rats had access to food and water *ad libitum.* The rats were acclimatized for a week prior to the start of the study.

Inducing Chronic Cocaine Dependence in Animal Models Using Conditioned Place Preference (CPP)

Chronic drug dependence was induced in the animal models according to a modified method of Martin *et al.* (2000) by means of a conditioned place preference paradigm. The conditioned place preference paradigm utilized a conditioned place preference (CPP) box. The CPP box consists of two compartments (15" x 15" x 15") that differed in both colour (white versus black) and light intensity (brightly lit versus dark and covered), separated by a neutral area. The rats were habituated to the conditioned place preference (CPP) box for 3 days by allowing them to run freely throughout the two compartments via the neutral area.

Rats in groups CC, CDND and CDNS were brought to a stage of drug dependence by use of the CPP box. The rats in groups ND and NS were used as controls and were brought through the CPP method, but instead using only vehicle (normal saline). This CPP paradigm occurred in three phases and was outlined as described below:

1. **Preconditioning Phase:** Each rat was placed in the neutral compartment and allowed to explore both compartments freely. During this stage, the time spent in each compartment was recorded out of a total 18 minutes and used in order to pair the compartments for association with either the drug or the vehicle

Table 1. Rat groups and their assigned diets

Group	Group Description
ND	Normal rats fed normal rat diet.
NS	Normal rats fed supplemented rat diet.
CC	Rats dependent on cocaine and continually given cocaine throughout the entire study period, fed normal rat diet.
CDND	Rats dependent on cocaine, cocaine use discontinued, fed normal rat diet.
CDNS	Rats dependent on cocaine, cocaine use discontinued, fed supplemented rat diet

(normal saline). The preferred compartment was paired to vehicle and the non-preferred compartment was paired with the cocaine hydrochloride solution.

2. **Conditioning Phase:** Rats were treated with alternate injections of cocaine hydrochloride (20 mg/kg in 1 cc. normal saline) or vehicle (normal saline) intra-peritoneally (i.p.) for 6 consecutive days. They were confined to the corresponding compartment immediately following injection using guillotine doors which matched the walls for 25 minutes. Drug was administered on days 1, 3 and 5 whereas the vehicle was administered on days 2, 4 and 6. Control subjects received vehicle every day. Following the drug pairing phase, rats in groups CC, CDND and CDNS were chronically administered cocaine at a dosage of 25 mg/kg in 1 cc. normal saline i.p. for 20 consecutive days. This occurred as stated above in the drug pairing phase. Control animals received vehicle every day and were treated in the same manner.
3. **Testing Phase:** This phase was carried out as described previously in the preconditioning phase. Free access was given to the rats of both compartments via the neutral chamber and the time spent in each compartment was recorded out of a total 18 minutes. This step occurred after every phase of the feeding trial, and weekly throughout the supplement administration phase.

Supplement Administration

The SA supplement formulation had three constituents and was introduced to the rats at dosages specified in its proprietary form based on body weights. Rat chow (PMI Nutrition Inc., Lab diet 5008) was milled to a powder to facilitate the homogenous mixing of the supplement constituents obtained from Getwele Foundation International (USA).

Rats in group CC were continually subjected to the drug throughout the entire course of the study at the same dosage and method as described previously. After the 20 day chronic drug administration period, rats in groups CDND and CDNS had their cocaine use discontinued and fed normal rat diet and supplemented rat diet respectively for a period of three months according to the SA supplement guidelines. Rats in group NS were fed supplemented rat diet henceforth. Other groups were continued with the normal rat diet.

Sample Collection

Blood was obtained from the rats after acclimatization (baseline), after chronic drug dependence, at the mid-point and the end of the supplement administration period. Serum was obtained for analyses following the centrifugation (x 5000 rpm for 5 minutes, Centrifuge 5402 (Eppendorf, Germany)) of the blood samples.

Analysis of Cocaine Metabolite Plasma Concentrations

Cocaine hydrochloride and its metabolites were determined by a direct competitive ELISA according to manufacturer's guidelines (Neogen Cooperation Cocaine/ Benzoylecgonine ELISA kit, 2006). The principle of the test was based on competition between a horseradish peroxidase enzyme conjugate and the cocaine metabolite that is present in the sample, with a limited number of specific binding sites on the microplate precoated with antibody. A colorimetric method was then employed to compare the absorbance readings of the samples with the standards using a microplate reader at 650 nm (ElX800 Universal Microplate Reader, Bio-Tek Instrument Inc., USA).

Liver Cytochrome P450 3A4 Isoenzyme Determination

Cytochrome P450 3A4 isoenzyme was determined in the liver utilizing a luminescent method following manufacturer's instructions in a kit from Promega, P450-Glo V9002 Assays (2011). The available cytochrome P450 isoenzymes present in the samples were reacted with a luminogenic cytochrome P450 substrate. The cytochrome P450 isoenzymes converted the substrate to luciferin, which in turn reacted with luciferase to produce light. The amount of light produced was directly proportional to the cytochrome P450 activity on the samples.

Statistical Analysis

The results obtained were expressed as mean value ± the standard error of the mean (SEM). The differences among all the groups at $p \leq 0.05$ were tested using the Analysis of the variance (ANOVA) test. To determine whether the significant difference lies among the groups, the Duncan's multiple range test was used to compare the means. All statistical analyses were done using the statistical program SPSS version 16 (2007).

RESULTS

Effect of Supplement Administration on Craving Indicated by Time Spent in the White Compartment of the CPP Box

Figures 1 – 4 shows the time spent in the white compartment of the conditioned place preference (CPP) box as an indicator of craving for cocaine at varying phases of the supplement administration period. There were no significant differences (p

≥ 0.05) in the time spent in the white compartment of the CPP box between any of the groups during the acclimatization period. Groups ND and NS maintained a time of between 4 to 6 minutes for the remainder of the study. These groups were the normal groups not exposed to cocaine. Groups CC, CDND and CDNS showed remarkable increase ($p \leq 0.05$) in time spent in the white compartment during the drug pairing exercise and after chronic cocaine treatment (Figure 1). Group CC was maintained on cocaine throughout the remainder of the study and fed normal diet which resulted in a time constant at just below 13 minutes throughout the remainder of the study, indicating sustained craving for cocaine.

Groups CDND and CDNS both had their cocaine use discontinued following the chronic cocaine administration period. Both groups showed signs of reduced craving for the drug given the moderate time spent in the white compartment ($p \leq 0.05$) of the CPP box at around week 5 of the supplement administration period. By week 6 (Figure 3), group CDNS spent significantly less time ($p \leq 0.05$) in the white compartment of the CPP box than that of group CDND. This trend continued

Figure 1. Time spent in white compartment of the CPP box (minutes) as an indicator of craving following chronic cocaine administration in rat model

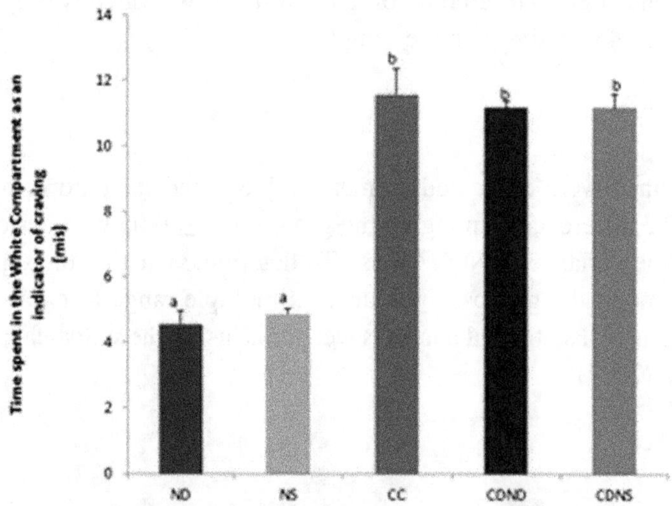

ND - *Normal rats fed normal rat diet*; NS - *Normal rats fed supplemented rat diet*; CC - *Rats dependent on cocaine and continually given cocaine throughout the entire study period, fed normal rat diet*; CDND - *Rats dependent on cocaine, cocaine use discontinued, fed normal rat diet*; CDNS - *Rats dependent on cocaine, cocaine use discontinued, fed supplemented rat diet.*
The values shown represent mean ± standard error of the mean.
Figures with different superscripts are significantly different ($p \leq 0.05$).

Figure 2. Time spent in the white compartment of the CPP box (minutes) as an indicator of craving one week following the start of nutritional supplement administration in rats

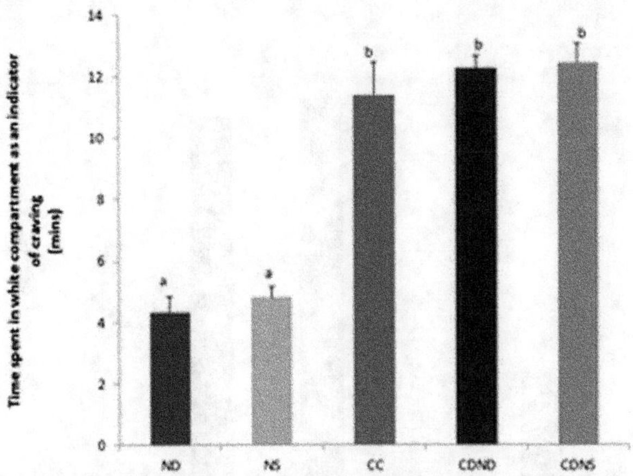

ND - *Normal rats fed normal rat diet*; NS - *Normal rats fed supplemented rat diet*; CC - *Rats dependent on cocaine and continually given cocaine throughout the entire study period, fed normal rat diet*; CDND - *Rats dependent on cocaine, cocaine use discontinued, fed normal rat diet*; CDNS - *Rats dependent on cocaine, cocaine use discontinued, fed supplemented rat diet.*
The values shown represent mean ± standard error of the mean.
Figures with different superscripts are significantly different ($p \leq 0.05$).

until the end of the supplement administration phase, where the time spent in the white compartment of the CPP box for groups CDND and CDNS was recorded at approximately 10 minutes and 6 minutes respectively.

Effect of Supplement Administration on Serum Cocaine Metabolite Levels

Table 2 illustrates the cocaine metabolite concentration in plasma of rats at varying intervals of the animal study. There was no cocaine metabolite observed in the groups not exposed to cocaine. Following chronic cocaine dependence, there was significant increase in the metabolite concentration in the groups of rats exposed to cocaine (CC, CDND and CDNS). Mid-way the supplement administration phase, there was a high level of cocaine metabolite in the circulating plasma in the group continually given cocaine throughout the entire study period. There was significant increase ($p \leq 0.05$) in the circulating cocaine metabolite concentrations in the groups that had their cocaine use discontinued and fed nutritionally supplemented diet in comparison to the group given normal diet.

Figure 3. Time spent in the white compartment of the CPP box (minutes) as an indicator of craving mod-way following the start of nutritional supplement administration period in experimental rat groups NS and CDNS

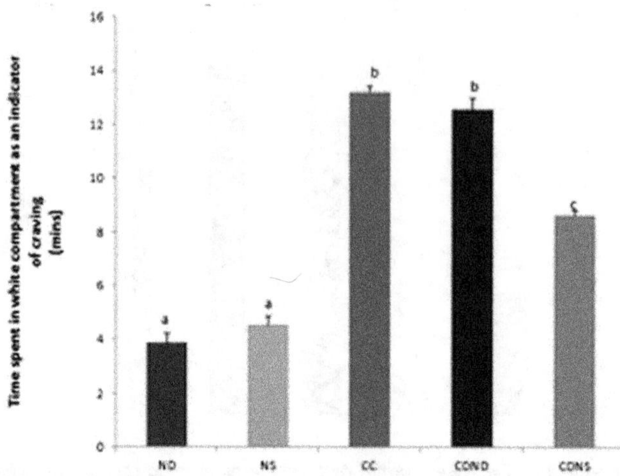

ND - *Normal rats fed normal rat diet*; NS - *Normal rats fed supplemented rat diet*; CC - *Rats dependent on cocaine and continually given cocaine throughout the entire study period, fed normal rat diet*; CDND - *Rats dependent on cocaine, cocaine use discontinued, fed normal rat diet*; CDNS - *Rats dependent on cocaine, cocaine use discontinued, fed supplemented rat diet.*
The values shown represent mean ± standard error of the mean.
Figures with different superscripts are significantly different ($p \leq 0.05$).

Table 2. The cocaine hydrochloride metabolite (benzoylecgonine) concentration in plasma of rats at varying intervals of the animal study

	Cocaine hydrochloride metabolite (benzoylecgonine) (ng/mL)			
	Phase of Animal Study			
Group	Baseline	After Chronic Cocaine Dependence	Mid-way Supplement Administration	End of Supplement Administration
ND	None detected	None detected	None detected	None detected
NS	None detected	None detected	None detected	None detected
CC	None detected	1090.32 ± 13.56^a	1141.20 ± 30.11^a	251.78 ± 31.95^b
CDND	None detected	1099.32 ± 25.45^a	203.89 ± 37.18^b	None detected
CDNS	None detected	1147.88 ± 31.41^a	361.83 ± 91.88^c	None detected

ND - *Normal rats fed normal rat diet*; NS - *Normal rats fed supplemented rat diet*; CC - *Rats dependent on cocaine and continually given cocaine throughout the entire study period, fed normal rat diet*; CDND - *Rats dependent on cocaine, cocaine use discontinued, fed normal rat diet*; CDNS - *Rats dependent on cocaine, cocaine use discontinued, fed supplemented rat diet.*
Results are expressed as mean ± standard error of the mean.
Figures with different letters are significantly different ($p \leq 0.05$).

Figure 4. Time spent in the white compartment of the CPP box (minutes) as an indicator of craving at the end of the nutritional supplement administration period in experimental rat groups NS and CDNS

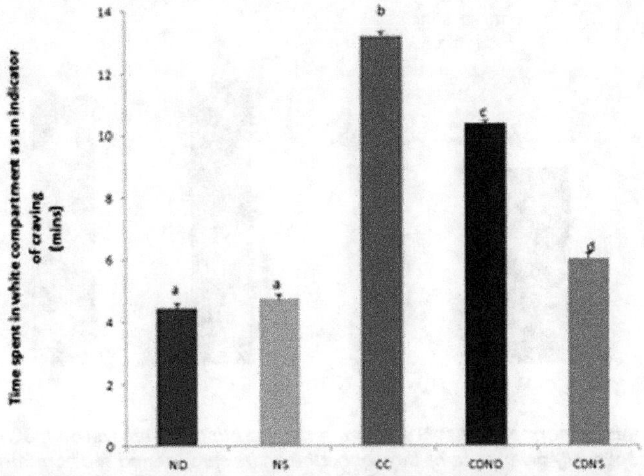

ND - *Normal rats fed normal rat diet*; NS - *Normal rats fed supplemented rat diet*; CC - *Rats dependent on cocaine and continually given cocaine throughout the entire study period, fed normal rat diet*; CDND - *Rats dependent on cocaine, cocaine use discontinued, fed normal rat diet*; CDNS - *Rats dependent on cocaine, cocaine use discontinued, fed supplemented rat diet.*
The values shown represent mean ± standard error of the mean.
Figures with different superscripts are significantly different ($p \leq 0.05$).

Cytochrome P450 3A4 Isoenzyme Liver Activity

Figure 5 illustrates cytochrome P450 3A4 isoenzyme activity in the liver at the end of the study period. Increased activity ($p \leq 0.05$) was observed in the group NS (fed the SA supplement formulation) when compared with the normal control group, ND. Groups CC, CDND and CDNS (the groups exposed to cocaine) exhibited increased enzyme activity ($p \leq 0.05$) where group CDNS in turn showed a significant increase ($p \leq 0.05$) in enzyme activity.

DISCUSSION

Conditioned Place Preference (CPP)

In this study rats with chronic cocaine dependence were developed using the conditioned place preference (CPP) paradigm. These rats were used as indicators of

Figure 5. CYP450 3A4 isoenzyme analysis following chronic cocaine exposure and subsequent treatment with SA supplements in rats

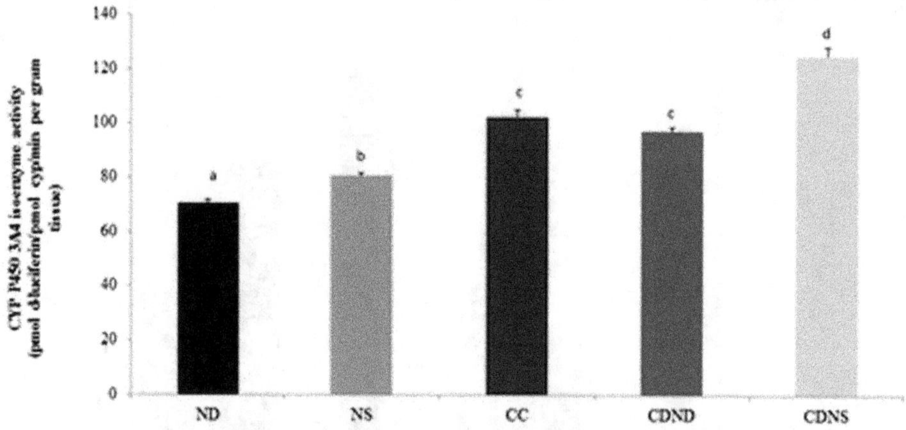

ND - *Normal rats fed normal rat diet*, NS - *Normal rats fed supplemented rat diet*, CC - *Rats dependent on cocaine and continually given cocaine throughout the entire study period, fed normal rat diet*, CDND - *Rats dependent on cocaine, cocaine use discontinued, fed normal rat diet*, CDNS - *Rats dependent on cocaine, cocaine use discontinued, fed supplemented rat diet*.
Figures represent mean ± standard error of the mean.
Figures with different letters are significantly different ($p \leq 0.05$).

craving, in order to determine the efficacy of the SA supplement with discontinued cocaine use, in order to elucidate the proposed mechanism of action of the supplements.

CPP has been developed and is used as one of the most effective experimental protocols to measure drug reward when using laboratory animals (Bardo & Bevins, 2000; Cunningham *et al.*, 2006) and has been used in numerous studies as a standard measure of craving in animal models (Carboni & Vacca, 2003; Tzschentke, 1998). The principle underlying the use of a CPP box is based on the choice for an otherwise 'unpreferred' compartment as an indication of craving. In this experiment, the pre-conditioning phase indicated that the 'preferred' compartment was the black, dark chamber in which the animal models spent the majority of the time naturally, as opposed to the white, brightly lit chamber which was the 'unpreferred' compartment, and where significantly less time was spent. The white, brightly lit chamber was paired with cocaine subsequently and a craving was noted by the significant increase in time spent in this once 'unpreferred' compartment. A marked craving for the drug was demonstrated by a significant increase in the amount of time spent in the white compartment after chronic drug administration in comparison to the periods before any cocaine was administered.

Other studies have consistently used this method where time spent in the chambers was indicative of craving for the drug (Maldonado & Rodriguez de Fonseca, 2002; Martin *et al.*, 2000; Sora *et al.*, 1998). Results indicate that there was significant craving produced for the drug prior to supplement administration in the groups that were exposed to cocaine chronically, namely groups CC, CDND and CDNS.

The effectiveness of the supplements on craving was tested in one group and compared with normal rat diet given to another group following chronic cocaine administration. The rats that were continually exposed to cocaine throughout the study consistently exhibited a craving for the drug throughout the study. At week 5 of the supplement administration period, group CDNS showed evidence of reduced craving for the drug as seen by a reduction in the time spent in the white compartment from 12.46 ± 0.18 minutes at the start of the administration to 11.38 ± 0.18 minutes. Over time, this value decreased until the end of the administration period. At week 13, the time spent in the white compartment was 6.05 ± 0.18 minutes, which was comparable to the group's time at acclimatization 5.98 ± 0.48 minutes. Group CDND exhibited a slight decrease in time spent in this compartment, but not as marked as group CDNS whereas group CC exhibited a continued indication of craving for the drug.

These results demonstrate that the administration of the supplements reduced craving of the drug significantly following discontinued cocaine use during the supplement administration period.

Cocaine Metabolism/Plasma Metabolites

Cocaine is metabolised in the body in numerous ways. The two primary means of metabolism incorporate the esterases which are found in various tissues and organs and the oxidative pathway, performed by the liver. Cocaine can release epinephrine and norepinephrine into the circulating system which may give way to glutathione depletion in the liver (James *et al.*, 1983; Register & Bartlett, 1954). Catecholamines seem to be elevated by some cocaine metabolites, which may lead to cocaine toxicity (Misra & Pontani, 1977; Williams *et al.*, 1977).

The primary cocaine metabolite, benzoylecgonine, was measured in the circulating plasma of rats throughout the entire study period. There was a high circulating plasma concentration of benzoylecgonine following chronic cocaine administration which was maintained over a period of time by the group continually administered cocaine throughout the study period. Cocaine as previously mentioned has been found to be stored in organs in the body when administered chronically (Bystrowska *et al.*, 2012; Colucci *et al.*, 2010). Circulating plasma concentrations are often an indicator of equilibrium between these organs operating as storage and the blood, which gives an indication of the concentrations remaining in the body (Chaing & Hawks, 1986; Rook *et al.*, 2006). The concentrations were significantly higher in the treatment group (CDNS) when compared to the untreated group (CDND). The higher equilibrium that appears to exist in the treatment group could possibly be an indicator of a gradual, slow release of the stored metabolites which may have contributed to better management of craving in the animal models.

Cytochrome P450 3A4 Isoenzyme Concentration

The cytochrome P450 system is one major system that carries out the metabolism of cocaine (Shuster *et al.*, 1983), particularly the cytochrome P450 3A4 isoenzyme activities which were determined at the end of the animal study using liver cells. Increased activity of the cytochrome P450 3A4 isoenzyme which is partly responsible for cocaine metabolism is suggestively indicative of increased metabolism of cocaine into its varying metabolites. The activities were significantly increased in the rats that were exposed to cocaine. This increased activity was further enhanced in the treated group (CDNS) which leads to the possible premise that the SA supplement formulation increases the breakdown of the cocaine into its metabolites and somehow provides a protective effect on the liver.

SUMMARY

The study was to investigate the effects of chronic cocaine administration and subsequent administration of nutritional supplements in reducing craving in this rat model. It was seen that the supplements reduced craving of cocaine significantly following discontinued cocaine in the animal model. The cytochrome P450 3A4 isoenzyme is partly responsible for cocaine metabolism and is suggestively indicative of increased metabolism of cocaine, which was further enhanced by the supplements. Blood plasma analysis showing higher benzoylecgonine equilibrium demonstrated that the supplements possibly have aided in removing the stored metabolites and this may have contributed to better management of craving in the rats.

ACKNOWLEDGMENT

The authors would like to thank The School of Graduate Studies and Research (The University of the West Indies, Mona), Environmental Health Foundation (Jamaica), Scientific Research Council (Jamaica) and Bureau of Standards (Jamaica) – the funding agencies for providing support.

REFERENCES

Affi, N. H., Abdel-Rahman, M. S., & Nassar, A. M. (1998). Effect of alcohol and/or cocaine on blood glutathione and the ultrastructure of the liver of pregnant CF-1 mice. *Toxicology Letters*, *98*(1), 1–12. doi:10.1016/S0378-4274(98)00040-X PMID:9776556

Bardo, M. T., & Bevins, R. (2000). Conditioned Place Preference: What does it add to our preclinical understanding of drug reward? *Psychopharmacology*, *153*(1), 31–43. doi:10.1007002130000569 PMID:11255927

Bates, C. K. (1988). Medical risks of cocaine use. *The Western Journal of Medicine*, *148*(4), 440. PMID:3291396

Boyer, C. S., & Peterson, D. R. (2005). Hepatic biochemical changes as a result of acute cocaine administration in the mouse. *Hepatology (Baltimore, Md.)*, *14*(6), 1209–1216. doi:10.1002/hep.1840140641 PMID:1959871

Bystrowska, B., Adamczyk, P., Moniczewski, A., Zaniewska, M., Fuxe, K., & Filip, M. (2012). LC/MS/MS evaluation of cocaine and its metabolites in different brain areas, peripheral organs and plasma in cocaine self-administering rats. *Pharmacological Reports*, *64*(1337), 1337–1349. doi:10.1016/S1734-1140(12)70931-3 PMID:23406744

Carboni, E., & Vacca, C. (2003). Conditioned place preference. In *Drugs of Abuse* (pp. 481–498). Humana Press.

Center for Substance Abuse Treatment. (1999). Brief Interventions and Brief Therapies for Substance Abuse. Rockville, MD: Substance Abuse and Mental Health Services Administration.

Chiang, C.N., & Hawks, R.L. (1986). Implications of drug levels in body fluids: Basic concepts. *Urine Testing for Drugs of Abuse, 73*, 62-83.

Colucci, A. P., Aventaggiato, L., Centrone, M., & Gagliano-Candela, R. (2010). Validation of an Extraction and Gas Chromatography-Mass Spectrometry Quantification Method for Cocaine, Methadone, and Morphine in Postmortem Adipose Tissue. *Journal of Analytical Toxicology*, *34*(6), 342–346. doi:10.1093/jat/34.6.342 PMID:20663287

Cunningham, C. L., Gremel, C. M., & Groblewski, P. A. (2006). Drug-induced conditioned place preference and aversion in mice. *Nature Protocols*, *1*(4), 1662–1670. doi:10.1038/nprot.2006.279 PMID:17487149

Espinoza, L. R., & Alamino, R. P. (2012). Cocaine-induced vasculitis: Clinical and immunological spectrum. *Current Rheumatology Reports*, *14*(6), 532–538. doi:10.100711926-012-0283-1 PMID:22875288

Garcia-Rodriguez, O., Secades-Villa, R., Higgins, S. T., Fernandez-Hermida, J. R., Carballo, J. L., Errasti Perez, J. M., & Diaz, S. A. (2009). Effects of voucher-based intervention on abstinence and retention in an outpatient treatment for cocaine addiction: A randomized controlled trial. *Experimental and Clinical Psychopharmacology*, *17*(3), 131–138. doi:10.1037/a0015963 PMID:19586227

Gutman, S. A. (2006, June). Why addiction has a chronic, relapsing course. The neurobiology of addiction: Implications for occupational therapy practice. *Occupational Therapy in Mental Health*, *22*(2), 1–29. doi:10.1300/J004v22n02_01

Hawks, R. L., Kopin, I. J., Colburn, R. W., & Thoa, N. B. (1974). Norcocaine: A pharmacologically active metabolite of cocaine found in brain. *Life Sciences*, *15*(12), 2189–2195. doi:10.1016/0024-3205(74)90035-6 PMID:4621015

Heard, K., Palmer, R., & Zahniser, N. R. (2008). Mechanisms of acute cocaine toxicity. *The Open Pharmacology Journal*, *2*(9), 70–78. doi:10.2174/1874143600802010070 PMID:19568322

Hu, X. (2007). Cocaine withdrawal and neuro-adaptations in ion channel function. *Molecular Neurobiology*, *35*(1), 95–112. doi:10.1007/BF02700626 PMID:17519508

James, R. C., Roberts, S. M., & Harbison, R. D. (1983). The perturbation of hepatic glutathione by α-adrenergic agonists. *Fundamental and Applied Toxicology*, *3*(4), 303–308. doi:10.1016/S0272-0590(83)80144-4 PMID:6138290

Kanel, G. C., Cassidy, W., Shuster, L., & Reynolds, T. B. (1989). Cocaine-induced liver cell injury: Comparison of morphological features in man and in experimental models. *Hepatology (Baltimore, Md.)*, *11*(4), 646–651. doi:10.1002/hep.1840110418 PMID:2328957

Kennedy, A. P., Gross, R. E., Whitfield, N., Drexler, K. P. G., & Kilts, C. D. (2012). A controlled trial of the adjunct use of D-cycloserine to facilitate cognitive behavioral therapy outcomes in a cocaine-dependent population. *Addictive Behaviors*, *37*(8), 900–907. doi:10.1016/j.addbeh.2012.03.008 PMID:22578380

Koob, G. F., & Le Moal, M. (2001). Drug addiction, dysregulation of reward, and allostasis. *Neuropsychopharmacology*, *24*(2), 97–129. doi:10.1016/S0893-133X(00)00195-0 PMID:11120394

Levine, S. R., Brust, J. C. M., Futrell, N., Ho, K.-L., Blake, D., Millikan, C. H., ... Welch, K. M. A. (1990). Cerebrovascular complications of the use of the Crack form of alkaloidal cocaine. *The New England Journal of Medicine*, *323*(11), 699–704. doi:10.1056/NEJM199009133231102 PMID:2388668

MacLachlan, P. L., & Hodge, H. C. (1938). The Influence of Cocaine Feeding on the Liver Lipids of the White Mouse. *The Journal of Biological Chemistry*, *104*, 721–726.

Maldonando, R., & Rodriguez de Fonseca, F. (2002). Cannabinoid Addiction: Behavioural Models and Neural Correlates. *The Journal of Neuroscience*, *22*(9), 3326–3331. doi:10.1523/JNEUROSCI.22-09-03326.2002 PMID:11978807

Martin, M., Ledent, C., Parmentier, M., Monaldonado, R., & Valverde, O. (2000). Cocaine, but not morphine, induces conditioned place preference and sensitization to locomotor responses in CB1 knockout mice. *The European Journal of Neuroscience*, *12*(11), 4038–4046. doi:10.1046/j.1460-9568.2000.00287.x PMID:11069600

Maude-Griffin, P. M., Hohenstein, J. M., Humfleet, G. L., Reilly, P. M., Tusel, D. J., & Hall, S. M. (1998). Superior efficacy of cognitive-behavioral therapy for urban crack cocaine abusers: Main and matching effects. *Journal of Consulting and Clinical Psychology*, *66*(5), 832–837. doi:10.1037/0022-006X.66.5.832 PMID:9803702

Misra, A. L., Nayak, P. K., Bloch, R., & Mule, S. J. (1975). Estimation and disposition of [3H] benzoylecgonine and pharmacological activity of some cocaine metabolites. *The Journal of Pharmacy and Pharmacology*, *27*(10), 784–786. doi:10.1111/j.2042-7158.1975.tb09404.x PMID:241797

Misra, A. L., & Pontani, R. B. (1977). Disposition and metabolism of psi-cocaine (dextrococaine) in the rat. *Drug Metabolism and Disposition: the Biological Fate of Chemicals*, *5*(6), 556–563. PMID:21779

National Institute on Drug Abuse. (2008). *Cocaine Drug Facts*. Retrieved from http://www.nida.nih.gov/pdf/infofacts/Cocaine08.pdf

Pasanen, M., Pellinen, P., Stenback, F., Juvonen, R. O., Raunio, H., & Pelkonen, O. (1995). The role of CYP enzymes in cocaine-induced liver damage. *Archives of Toxicology*, *69*(5), 287–290. doi:10.1007002040050172 PMID:7654131

Rebec, G. V. (2012). *Cocaine and Amphetamines in:eLS*. Chichester, UK: John Wiley & Sons Ltd. doi:10.1002/9780470015902.a0000042.pub3

Register, U. D., & Bartlett, R. G. (1954). Relationship of adrenalin to tissue sulfhydryl compounds. *Science*, *120*(3107), 109–110. doi:10.1126cience.120.3107.109 PMID:13178662

Robinson, T.E. & Berridge, K.C. (2000). The psychology and neurobiology of addiction: an incentive-sensitization view. *Addiction*, *95*(8s2), 91-117.

Rook, E. J., Huitema, A. D. R., Mv Ree, J., & Beijnen, J. H. (2006). Pharmacokinetics and pharmacokinetic variability of heroin and its metabolites: Review of the literature. *Current Clinical Pharmacology*, *1*(1), 109–118. doi:10.2174/157488406775268219 PMID:18666382

Shuster, L., Eleanor, C., & Sudhakar, S. W. (1983). Metabolism of cocaine and norcocaine to N-hydroxynorcocaine. *Biochemical Pharmacology*, *32*(20), 3045–3051. doi:10.1016/0006-2952(83)90247-2 PMID:6639673

Sora, I., Wichems, C., Takahashi, N., Li, X., Zeng, Z., Revay, R., ... Uhl, G. R. (1998). Cocaine reward models: Conditioned place preference can be established in dopamine- and in serotonin-transporter knockout mice. *Proceedings of the National Academy of Sciences of the United States of America*, *95*(13), 7699–7704. doi:10.1073/pnas.95.13.7699 PMID:9636213

Thompson, M. L., Shuster, L., & Shaw, K. (1979). Cocaine-induced hepatic necrosis in mice—The role of cocaine metabolism. *Biochemical Pharmacology*, *28*(15), 2389–2395. doi:10.1016/0006-2952(79)90705-6 PMID:497018

Tsai, L. and Zager, D. (2005). Medical Treatment in autism. *Autism spectrum disorders: Identification, education and treatment*, 395-492.

Tzschentke, T. M. (1998). Measuring reward with the conditioned place preference paradigm: A comprehensive review of drug effects, recent progress and new issues. *Progress in Neurobiology*, *56*(6), 613–672. doi:10.1016/S0301-0082(98)00060-4 PMID:9871940

UNODC. (2010). *United Nations Office on Drugs and Crime. World Drug Report 2010*. United Nations Publication.

Volkow, N.D. (2009). What Is Cocaine? In *Cocaine: Abuse and Addiction.* NIDA Research Report Series 3-7.

Weiss, R. D., Ostacher, M. J., Otto, M. W., Calabrese, J. R., Fossey, M., Wisniewski, S. R., ... Sachs, G. S. (2005). Does recovery from substance use disorder matter in patients with bipolar disorder? *The Journal of Clinical Psychiatry*, *66*(6), 730–735. doi:10.4088/JCP.v66n0609 PMID:15960566

Williams, J., & Wasserberger, J. (2006). Crack cocaine causing fatal vasoconstriction of the aorta. *The Journal of Emergency Medicine*, *31*(2), 181–184. doi:10.1016/j.jemermed.2005.08.021 PMID:17044582

Chapter 2
From Einstein Theories to Least Action Principle, a Relativistic Error of a Limit Case of Classical Mechanics:
Matching With Measurements of Light Propagation

Alessandro Massaro
Dyrecta Lab s.r.l., Italy

ABSTRACT

The authors prove that the standard least action principle implies a more general form of the same principle by which they can state generalized motion equation including the classical Euler equation as a particular case. This form is based on an observation regarding the last action principle about the limit case in the classical approach using symmetry violations. Furthermore, the well-known first integrals of the classical Euler equations become only approximate first integrals. The authors also prove a generalization of the fundamental lemma of the calculus of variation and they consider the application in electromagnetism. This chapter is an enhanced version of a published work. It proves the existence of particular relativistic error condition in classical mechanics, potentially significant on experiments of light propagation in matters. The work includes a discussion of applications potentially correlated with the found particle motion error condition.

DOI: 10.4018/978-1-5225-5406-6.ch002

INTRODUCTION

The least action principle is widely applicable in any field of physics and is particularly fruitful in classical mechanics (Goldstein et al., 2008). It is well known that almost all the physical equations can be derived by such a principle, so that a critical analysis about this principle is certainly useful.

The aim of this paper is to prove that its standard form implies a generalized variational principle by which we can derive generalized Euler equations: these latter are more general than the classical Euler equation and include them as a special case.

In stating our results we make use of a generalization of the fundamental lemma of the calculus of variations.

An attempt is made to interpret as a relativistic correction the difference between the generalized equations and the classical equations. Generally the new equations lead to approximate first integrals which become exact only in limiting case of the classical equations. The results of this analysis can be extended to classical field theory and to Relativistic physics and probably to Quantum mechanics and Quantum field theory (Massaro, 2011). Another application of this theory is in electron optics and, in particular, in the evaluation of the light velocity in a liquid. The chapter is an enhanced version of a work (Massaro & Massaro, 2011), which highlights the particular error condition of an analytical solution in classical mechanics theory matching with physical phenomena involving light propagation in a liquid. After an enhanced background discussion some aspect concerning the performed research and the new topics of research about relativistic error will be commented on.

BACKGROUND

Our recently funded project re-analyzed analytic mechanics structure of quantum mechanics and relativity space-time (Xinzhong, 2016). The topic is very important in order to "trace" accurately the analytical solutions from classical mechanics to quantum mechanics. We discuss here a short history of analytical features in order to understand different aspects of mathematical solutions of dynamics and thermodynamics and light propagation principles. Approximately in 1620 W.R. Snell studied the refraction of light ($n_1 \sin\alpha = n_2 \sin\beta$), by providing important aspects about ray optics. Successively Pierre de Fermat in 1662 studied deeply the rectilinear propagation, the reflection and the refraction of light, thus also defining the basic principle that when light pass through the interface of two kinds of medium, whether reflection or refraction, light also uses shortest path. In 1725 John Bernoulli studied the concept of balancing forces and energies affirming that in the balance of all forces, the sum of positive energy is always equal to the sum of the negative energy. In 1743 d' Alambert studied the effects of the Inertia force. In 1744 Pierre Louis Moreau De Maupertius discussed that the last action principle was suitable for collision problems. In the same year Lehonard Euler performed important researches about statics, particle trajectory curvatures, and multi-particle system. In the 1788 J.L. Lagrange started from the last action principle and took into account the true path and the variational path having the same energy thus defining the universal equation of d'Alambert-Lagrange. In the 1826 W.R. Hamilton analyzed deeply the Fermat principle in optics and kinetic energy of particles in motion, defining a new function H of the total energy of a system. In 1837 M. H. Jacobi eliminated Lagrange time variable in the last action principle by energy integral. In 1883-1884 H. L. F. Helmholtz defined relationships involving generalized force of phenomenon in mechanics, electrodynamics, electromechanics and thermodynamics. In the 1894 H. R. Hertz expanded Newton's inertia law to a particles' system without active forces. In 1913 Albert Einstein introduced the outline of general relativity on the theory of gravity: through classical mechanics and special relativity the scientist defined the equation of motion in the gravitational field. In 1918 E. Schrodinger focused his attention on relation between Einstein and Hertz. In 1911Arnord Sommerfeld defined the last action principle by means of the Lagrange function L of an electron to integrate over a time t representing the duration of the radiation. In 1923 Louis Vector de Broglie studied the wavefront of an object in motion for radiation and light quantum diffraction. In 1926 E. Schrodinger defined the Schrodinger wave equation (wavefunctions in presence of forces). In 1975 important studies have been performed on quantum mechanical effects causing black holes and emitting particles (Hawking, 1975).

In 1993 J. D. Norton re-discussed the principle of general covariance of Einstein's theory as a physically vacuous principle. (Norton, 1993). In 1995 D. F. Styer analyzed misconceptions concerning energy eigenstates (Styer, 1995).

Quantum Mechanics in three dimensions has been explained in a clear way more recently by other authors (Serway et al., 2005) enhancing defined applications of mathematical models. Some researchers studied the reconciliation to one law of principle of natural selection and principle of least action (Kaila & Annila, 2008). In the recent 2014, some authors showed some possible results of the history of physics of the least action and conservation of energy and concluded that "*the conservation of energy in time is a well-defined concept even in quantum mechanics*" (Oliveira, 2014).

THE ACTION FUNCTIONAL IN DIFERENT SETS OF ADMISSIBLE FUNCTIONS

We provide in this section and in the following ones some mathematical passages useful to determine the relativistic error condition.

Let us consider the action functional for a single particle moving along x-axis

$$S[x]=\int_{t_1}^{t_2} L(t,x,\dot{x})dt \tag{1}$$

where $L(t,x,\dot{x})$ is the Lagrangian. If x(t) is the real physical curve described by the particle (trajectory), such that $x(t_1)=a$, $x(t_2)=b$, we consider the functional (1) in the following sets of admissible functions

$$\begin{aligned}
\Gamma_1 &= \{x_1(t)=x(t)+h(t) \mid x(t)\in D_1(t_1,t_2), h(t)\in D_1(t_1,t_2), h(t_1)=0, h(t_2)=0\} \\
\Gamma_2 &= \{x_2(t,\alpha)=x(t)+\alpha h_0(t) \mid x(t)\in D_1(t_1,t_2), h_0(t)\in D_1(t_1,t_2), h_0(t_1)=0, h_0(t_2)=0, |\alpha|\le \alpha_0\} \\
\Gamma_3 &= \{x_3(t,\alpha)=x(t)+\alpha h_0(t)+\alpha^2 h_1(t) \mid x(t)\in D_1(t_1,t_2), h_0(t)\in D_1(t_1,t_2), h_1(t)\in D_1(t_1,t_2), \\
&\qquad h_0(t_1)=h_0(t_2)=0, h_1(t_1)\neq 0, h_1(t_2)\neq 0, \int_{t_1}^{t_2} h_1(t)dt\neq 0, |\alpha|\le\alpha_0\}
\end{aligned} \tag{2}$$

Here $D_1(t_1,t_2)$ is the space of all functions defined in the interval $[t_1,t_2]$ which are continuous and have continuous first derivatives; $h(t)$ is an arbitrary variable function, $h_0(t)$ and $h_1(t)$ are arbitrary but fixed functions; α is a real variable parameter. Figure 1 reports an example of x_2-trajectory.

The increment of action functional corresponding to the increment h(t) of x(t) is written as

$$\Delta S[x;h]=S[x+h]-S[x]=\delta S[x;h]+\varepsilon \,||\, h \,||\tag{3}$$

where $\delta S[x;h]$ is the differential of the functional and $\varepsilon\to 0$ for $||h||\to 0$. The last action principle states that in the set Γ_1 we have:

$$\delta S^{(1)}[x;h]=0 \tag{4}$$

Figure 1. Example of the trajectory $x_2(t,\alpha)$ from point A to point B

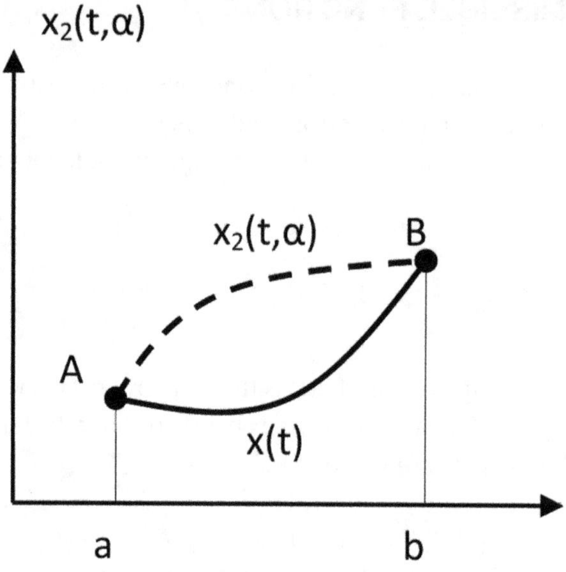

We now observe that:

$(\delta S^{(1)}[x]=0)\rightarrow(\delta S^{(2)}[x]=0)\rightarrow(\delta S^{(3)}[x]=0)$

In fact an explicit form of the formula (4) can be derived by using the Taylor's theorem. So we have

$$\begin{aligned}\Delta S[x;h] &= \int_{t_1}^{t_2} L(t,x+h,\dot{x}+\dot{h})dt - \int_{t_1}^{t_2} L(t,x,\dot{x})dt = \\ &= \int_{t_1}^{t_2} [L(t,x+h,\dot{x}+\dot{h}) - L(t,x,\dot{x})]dt = \\ &= \int_{t_1}^{t_2} [L_x(t,x,\dot{x})h + L_{\dot{x}}(t,x,\dot{x})\dot{h}]dt + \dots.\end{aligned} \tag{5}$$

where the subscript denotes partial derivatives with respect to the corresponding arguments, and the dots denote terms of order higher than 1 relative to h and $\dot{h}$ In this manner the formula (4) can be written as

$$\delta S^{(1)}[x;h] = \int_{t_1}^{t_2} [L_x(t,x,\dot{x})h + L_{\dot{x}}(t,x,\dot{x})\dot{h}]dt = 0 \tag{6}$$

By considering the action functional in the set Γ_2 we have

$$\begin{aligned}\delta S^{(2)}[x;\alpha h_0] &= \alpha \int_{t_1}^{t_2} [L_x(t,x,\dot{x})h_0 + L_{\dot{x}}(t,x,\dot{x})\dot{h}_0]dt = \\ &= \alpha \delta S^{(1)}[x;h_0] = 0\end{aligned} \tag{7}$$

So we have

$$(\delta S^{(1)}[x] = 0) \rightarrow (\delta S^{(2)}[x] = 0) \tag{8}$$

We can also prove the implication

$$(\delta S^{(2)}[x] = 0) \rightarrow (\delta S^{(3)}[x] = 0) \tag{9}$$

by showing that (Gelfand et al., Prentice-Hall, 1963)

$$\frac{\delta S^{(3)}[x;h]}{||h||} \xrightarrow[||h|| \rightarrow 0]{} 0 \tag{10}$$

In fact, by considering the action functional in the set Γ_3, we have

$$\begin{aligned}\delta S^{(3)}[x;\alpha h_0 + \alpha^2 h_1] &= \alpha \int_{t_1}^{t_2} [L_x(t,x,\dot{x})h_0 + L_{\dot{x}}(t,x,\dot{x})\dot{h}_0]dt + \\ &+ \alpha^2 \int_{t_1}^{t_2} [L_x(t,x,\dot{x})h_1 + L_{\dot{x}}(t,x,\dot{x})\dot{h}_1]dt +\end{aligned} \tag{11}$$

and therefore

$$\frac{\delta S^{(3)}[x;h]}{\|h\|} = \frac{\alpha \delta S^{(2)}[x;h_0]}{|\alpha| \; \|h_0 + \alpha h_1\|} + \frac{\alpha^2 \int_{t_1}^{t_2} [L_x(t,x,\dot{x})h_1 + L_{\dot{x}}(t,x,\dot{x})\dot{h}_1]dt}{|\alpha| \; \|h_0 + \alpha h_1\|} \xrightarrow[\alpha \to 0]{} 0 \tag{12}$$

Then we have $\delta S^{(3)}[x]=0$ being $\delta S^{(2)}[x]=0$. So we have proved the chain of implications $(\delta S^{(1)}[x]=0) \rightarrow (\delta S^{(2)}[x]=0) \rightarrow (\delta S^{(3)}[x]=0)$.

GENERALIZATION OF THE FUNDAMENTAL LEMMA OF THE CALCULUS OF VARIATIONS AND GENERALIZED EULER EQUATION

In this section we discuss the fundamental lemma of the calculus of variations and the relationship of the generalized Euler Equation.

By using a lemma of the calculus of variation (Gelfand et al., 1963) the classical Euler equation

$$\mathrm{L}_{\mathrm{x}} - \frac{d}{dt}\mathrm{L}_{\dot{\mathrm{x}}} = 0 \tag{13}$$

can be derived by the condition (6) and (7). But the condition $\delta S^{(3)}[x]=0$ contains more information. In fact we have

$$\delta S^{(3)}[x;h] = \int_{t_1}^{t_2} [L_x h + L_{\dot{x}} \dot{h}]dt = \int_{t_1}^{t_2} L_x h dt + [L_{\dot{x}} h]_{t_1}^{t_2} - \int_{t_1}^{t_2} \left(\frac{d}{dt} L_{\dot{x}} \right) h dt =$$
$$= \int_{t_1}^{t_2} \left[L_x - \frac{d}{dt} L_{\dot{x}} \right] h dt + \alpha^2 [L_{\dot{x}}|_{t=t_2} h_1(t_2) - L_{\dot{x}}|_{t=t_1} h_1(t_1)] = 0 \tag{14}$$

where $h(t)=\alpha h_0(t)+ \alpha^2 h_1(t)$.

By setting

$$\phi(t)=\mathrm{L}_x - \frac{d}{dt}\mathrm{L}_{\dot{x}} \tag{15}$$

$$c_{h_1} = \mathrm{L}_{\dot{x}}\Big|_{t=t_2} h_1(t_2) - \mathrm{L}_{\dot{x}}\Big|_{t=t_1} h_1(t_1) \tag{16}$$

we can write Equation (14) as

$$\alpha\int_{t_1}^{t_2}\phi(t)h_0(t)dt + \alpha^2\int_{t_1}^{t_2}\phi(t)h_1(t)dt + \alpha^2 c_{h_1} = 0, \quad |\alpha| \leq \alpha_0 \tag{17}$$

Hence we must have

$$\int_{t_1}^{t_2}\phi(t)h_0(t)dt = 0 \tag{18}$$

$$\int_{t_1}^{t_2}\phi(t)h_1(t)dt = -c_{h_1} \tag{19}$$

with arbitrary $h_0(t)$and $h_1(t)$.

By using the fundamental lemma of the calculus of variations (Landau, Addison Wesley Publishing Company,1959) equation (18) gives again the classical Euler equation $\phi(t)=0$. Equation (19) can be written as

$$\int_{t_1}^{t_2}[\phi(t) + \chi(h_1)]h_1(t)dt = 0 \tag{20}$$

where we have introduced the functional

$$\chi(h_1) = c_{h_1} \Big/ \int_{t_1}^{t_2} h_1(t)dt \qquad (21)$$

We can now establish the following generalized lemma of the calculus of variations: if $\phi(t)$ is continuous in $[t_1, t_2]$, and if equation (20) is satisfied for every function $h_1(t) \in C(t_1, t_2)$ such that $h_1(t) \neq 0$, $h_1(t_2) \neq 0$, being non zero the integral of equation (21), then we obtain

$$G(t, h_1) = \phi(t) + \chi(h_1) \equiv 0 \qquad (22)$$

for all $t \in [t_1, t_2]$ and all $h_1(t) \in C(t_1, t_2)$ ($C(t_1, t_2)$ is the space of continuous functions on the interval $[t_1, t_2]$)

Proof: Suppose that $\exists\ \bar{t} \in [t_1, t_2]$ and $\exists\ \bar{h}_1(t) \in C(t_1, t_2)$ such that $G(\bar{t}, \bar{h}_1) = \phi(\bar{t}) + \chi(\bar{h}_1)$ is non zero, say positive, at the point $\bar{t}$, being $\chi(\bar{h}_1)$ some constant. Then the function $G(\bar{t}, \bar{h}_1)$ is also positive in some interval $[\bar{t} - \delta, \bar{t} + \delta]$ contained in $[t_1, t_2]$.

If we set

$$h_1(t) = \begin{cases} \bar{h}_1(t_1) - \bar{h}_1(t_1)(t - t_1)/\delta' & , \quad t_1 \le t \le t_1 + \delta' \\ \lambda(t - \bar{t} + \delta)(\bar{t} + \delta - t) & , \quad \bar{t} - \delta \le t \le \bar{t} + \delta \\ \bar{h}_1(t_2) - \bar{h}_1(t_2)(t - t_2)/\delta' & , \quad \bar{t}_2 - \delta' \le t \le t_2 \\ 0 & , \quad \text{otherwise} \end{cases} \qquad (23)$$

we have $h_1(t_1) = \bar{h}_1(t_1)$, $h_1(t_2) = \bar{h}_1(t_2)$ and then $c_{h_1} = c_{\bar{h}_1}$.

In Figure 2 is shown the qualitative plot of the function (23) in the considered interval.

For $\delta' \to 0$ we can choose the parameter λ in such a way that we have

$$.\chi(h_1) = c_{h_1} \Big/ \int_{t_1}^{t_2} h_1(t)dt \ge \chi(\bar{h}_1) = c_{h_1} \Big/ \int_{t_1}^{t_2} \bar{h}_1(t)dt\,. \qquad (24)$$

In this case we obtain

Figure 2. Qualitative plot of the testing function $h_1(t)$

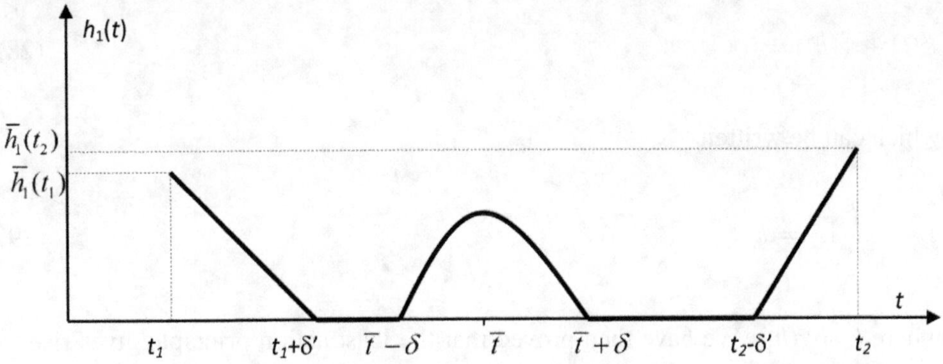

$$\int_{t_1}^{t_2}[\phi(t)+\chi(h_1)]h_1(t)dt=\int_{t_1}^{t_1+\delta'}[\phi(t)+\chi(h_1)]h_1(t)dt+\int_{t_2-\delta'}^{t_2}[\phi(t)+\chi(h_1)]h_1(t)dt$$
$$+\int_{\bar{t}-\delta}^{\bar{t}+\delta}[\phi(t)+\chi(h_1)]h_1(t)dt \quad (25)$$

For δ'→0 we can omit the first and the second integral on the right hand side of this equation and we obtain the evaluation

$$\int_{t_1}^{t_2}[\phi(t)+\chi(h_1)]h_1(t)dt\simeq\int_{\bar{t}-\delta}^{\bar{t}+\delta}[\phi(t)+\chi(h_1)]h_1(t)dt\neq 0 \quad (26)$$

This contradiction proves the lemma.

Remark: The lemma still holds if we replace $C(t_1,t_2)$ by $D_1(t_1,t_2)$. To see this we use the same proof with

$$h_1(t)=\begin{cases}\bar{h}_1(t_1)(t-t_1-\delta')^2/\delta'^2 & , \quad t_1\leq t\leq t_1+\delta' \\ \lambda[(t-\bar{t}+\delta)(\bar{t}+\delta-t)]^2 & , \quad \bar{t}-\delta\leq t\leq \bar{t}+\delta \\ \bar{h}_1(t_2)(t-t_2-\delta')^2/\delta'^2 & , \quad t_2-\delta'\leq t\leq t_2 \\ 0 & , \quad \text{otherwise}\end{cases} \quad (27)$$

Now equation (20) tells us that we also have the generalized Euler equation

$$\phi(t)+\chi(h_1)=0 \tag{28}$$

which can be written

$$L_x-\frac{d}{dt}L_{\dot{x}}=k_h \tag{29}$$

where k_h=-$\chi(h_1)$. We have then proved that the least action principle gives rise to the classical Euler equation or to the generalized Euler equation which becomes the classical equation in the case k_h=0.

In any case, to fit the experimental data we must assume very small values for k_h.

AN EVALUATION FOR KH AS RELATIVISTIC AND NON-RELATIVISTIC ERROR OF A MOVING PARTICLE

The constant k_h appearing in the generalized Euler equation (29) is arbitrary but we can obtain an evaluation for this constant by substituting it by a function $k_h(\dot{x},\ddot{x})$ and assuming that the deviation of equation (29) from the classical equation (13) is simply due to a relativistic correction. This function can be determined as follows. For a single free particle moving along x-axis the non-relativistic and relativistic Lagrangian are respectively (Bergmann, 1942)

$$L^{(NR)}=\frac{1}{2}m\dot{x}^2 \tag{30}$$

$$\begin{aligned}L^{(R)}&=-mc^2\left(1-\frac{\dot{x}^2}{c^2}\right)^{1/2}=-mc^2\left[1-\frac{1}{2}\frac{\dot{x}^2}{c^2}-\frac{1}{8}\frac{\dot{x}^4}{c^4}+\ldots\right]=\\&=-mc^2+\frac{1}{2}m\dot{x}^2+\frac{1}{8}m\frac{\dot{x}^4}{c^2}+\ldots\end{aligned} \tag{31}$$

The generalized Euler equation are

$$-\frac{d}{dt}\mathrm{L}_{\dot{x}}^{(\mathrm{NR})} = -m\ddot{x} = k_h^{(\mathrm{NR})} \tag{32}$$

$$-\frac{d}{dt}\mathrm{L}_{\dot{x}}^{(\mathrm{R})} = -\frac{d}{dt}\mathrm{L}_{\dot{x}}^{(\mathrm{NR})} - \frac{1}{2}m\frac{d}{dt}\frac{\dot{x}^3}{c^2} \tag{33}$$

From equation (32) and equation (33) we find

$$k_h^{(\mathrm{NR})} = k_h^{(\mathrm{R})} + \frac{1}{2}m\frac{d}{dt}\frac{\dot{x}^3}{c^2} = k_h^{(\mathrm{R})} + \frac{3}{2}m\frac{\dot{x}^2}{c^2}\ddot{x} \tag{34}$$

We observe that $k_h^{(\mathrm{NR})}$ and $k_h^{(\mathrm{R})}$ may represent the relativistic and non relativistic error, respectively, for a measure of a moving particle.

If we assume $k_h^{(\mathrm{R})}=0$, i.e. assume the exact classical Euler in special Relativity, we have

$$k_h^{(\mathrm{NR})} = \frac{3}{2}m\frac{\dot{x}^2}{c^2}\ddot{x} \tag{35}$$

so that the non-relativistic generalized Euler equation is in this case

$$-\ddot{x} = \frac{3}{2}\frac{\dot{x}^2}{c^2}\ddot{x} \tag{36}$$

and has the classical solution

$$\dot{x} = \dot{x}_0 = const \tag{37}$$

We can show that also for $k_h^{(\mathrm{R})}\neq 0$ we have the classical solution (37). In fact we have by definition

$$k_h^{(R)} = -\frac{d}{dt}\mathrm{L}_{\dot{x}}^{(R)} = -\frac{d}{dt}\frac{m\dot{x}}{\left(1-\dfrac{\dot{x}^2}{c^2}\right)^{1/2}} = -m\ddot{x}\left(1-\frac{\dot{x}^2}{c^2}\right)^{-1/2} - m\ddot{x}\frac{\dot{x}^2}{c^2}\left(1-\frac{\dot{x}^2}{c^2}\right)^{-3/2} \simeq$$

$$\simeq -m\ddot{x}\left(1+\frac{1}{2}\frac{\dot{x}^2}{c^2}\right) - m\ddot{x}\frac{\dot{x}^2}{c^2}\left(1+\frac{3}{2}\frac{\dot{x}^2}{c^2}\right) = -m\ddot{x} - \frac{3}{2}m\ddot{x}\frac{\dot{x}^2}{c^2} - \frac{3}{2}m\ddot{x}\frac{\dot{x}^4}{c^4} \tag{38}$$

so in this case we have by equation (34)

$$k_h^{(NR)} = -m\ddot{x} - \frac{3}{2}m\ddot{x}\frac{\dot{x}^4}{c^4} \tag{39}$$

The non-relativistic generalized Euler equation is then

$$\ddot{x}\frac{\dot{x}^4}{c^4} = 0 \tag{40}$$

which has the solution (37).

Nevertheless it is important to note that if a force, e.g. a constant force, is acting upon the particle the motion is a bit different from the classical motion. In fact in this case the non-relativistic Lagrangian is

$$\mathrm{L}^{(NR)} = \frac{1}{2}m\dot{x}^2 - V(x) \tag{41}$$

with F=-dV/dx=const.

By assuming (35), the non-relativistic generalized Euler equation is now

$$F - m\ddot{x} = \frac{3}{2}m\frac{\dot{x}^2}{c^2}\ddot{x} \tag{42}$$

and by integrating this equation e.g. with $\dot{x}_0 = 0$, we obtain

$$\dot{x}\left(1+\frac{1}{2}\frac{\dot{x}^2}{c^2}\right)=\frac{F}{m}t \tag{43}$$

CANONICAL FORM OG GENERALIZED EULER EQUATION

For a system consisting of a finite number of particles we have an action functional

$$S[x_1, x_2, ..., x_n] = \int_{t_1}^{t_2} L(t, x_1, x_2, ..., x_n, \dot{x}_1, \dot{x}_2, ..., \dot{x}_n) dt \tag{44}$$

and a system of n second-order generalized Euler equations

$$\mathrm{L}_{x_i} - \frac{d}{dt}\mathrm{L}_{\dot{x}_i} = k_{h_i} \qquad (i = 1, 2, ..., n) \tag{45}$$

These equations can be reduced in the standard way to a system of 2n first-order differential equations. By regarding $\dot{x}_1, \dot{x}_2, ..., \dot{x}_n$ as independent function of $x_1, x_2, ..., x_n$ we can write equation (45) in the form

$$\frac{dx_i}{dt} = \dot{x}_i \quad , \quad \mathrm{L}_{x_i} - \frac{d}{dt}\mathrm{L}_{\dot{x}_i} = k_{h_i} \qquad (i = 1, 2, ..., n) \tag{46}$$

where $x_1, x_2, ..., x_n, \dot{x}_1, \dot{x}_2, ..., \dot{x}_n$ are 2n unknown functions, and t the independent variable.

Let

$$p_i = \mathrm{L}_{\dot{x}_i} \tag{47}$$

and suppose that the Jacobian

$$\frac{\partial(p_1, p_2, ..., p_n)}{\partial(\dot{x}_1, \dot{x}_1, ..., \dot{x}_n)} = \det \| \mathrm{L}_{\dot{x}_i \dot{x}_k} \| \tag{48}$$

is non zero.

We can write $\dot{x}_1, \dot{x}_2, ..., \dot{x}_n$ as functions of the canonical variables $t, x_1, x_2, ..., x_n, p_1, p_2, ..., p_n$ and introduce the Hamiltonian $H(t, x_1, x_2, ..., x_n, p_1, p_2, ..., p_n)$ by the formula

$$\mathrm{H} = -\mathrm{L} + \sum_{i=1}^{n} \dot{x}_i p_i \tag{49}$$

To obtain the canonical form of generalized Euler equation we express L_{xi} in terms of H_{xi}, by evaluating the differential of the function H

$$d\mathrm{H} = -d\mathrm{L} + \sum_{i=1}^{n} p_i d\dot{x}_i + \sum_{i=1}^{n} \dot{x}_i dp_i = -\frac{\partial \mathrm{L}}{\partial t} dt - \sum_{i=1}^{n} \frac{\partial \mathrm{L}}{\partial x_i} dx_i - \sum_{i=1}^{n} \frac{\partial \mathrm{L}}{\partial \dot{x}_i} d\dot{x}_i + \\ + \sum_{i=1}^{n} p_i d\dot{x}_i + \sum_{i=1}^{n} \dot{x}_i dp_i = \\ = \frac{\partial \mathrm{H}}{\partial t} + \sum_{i=1}^{n} \frac{\partial \mathrm{H}}{\partial x_i} dx_i + \sum_{i=1}^{n} \frac{\partial \mathrm{H}}{\partial p_i} dp_i \tag{50}$$

Therefore

$$\frac{\partial \mathrm{H}}{\partial t} = \frac{\partial \mathrm{L}}{\partial t}, \ \frac{\partial \mathrm{H}}{\partial x_i} = -\frac{\partial \mathrm{L}}{\partial x_i}, \frac{\partial \mathrm{H}}{\partial p_i} = \dot{x}_i \tag{51}$$

i.e., by introducing the generalized Euler equations (46) we obtain their canonical form

$$\frac{dx_i}{dt} = \frac{\partial \mathrm{H}}{\partial p_i} \quad , \quad \frac{dp_i}{dt} = \frac{\partial \mathrm{H}}{\partial x_i} - k_{h_i} \tag{52}$$

APPROXIMATE FIRST INTEGRALS OF GENERALIZED EULER EQUATION

We call first integral of a system of differential equations every function which has a constant value along each integral curve of the system. Now we can show that the generalized Euler equations (52) have only approximate first integrals which become exact in the limit k_{hi} =0. If the Lagrangian L does not depend on t explicitly, i.e. is of the form $L(x_1, x_2, ..., x_n, \dot{x}_1, \dot{x}_2, ..., \dot{x}_n)$, then Hamiltonian (49) also does not depend on t explicitly but we have

$$\frac{dH}{dt} = \sum_{i=1}^{n}\left(\frac{\partial H}{\partial x_i}\frac{dx_i}{dt} + \frac{\partial H}{\partial p_i}\frac{dp_i}{dt}\right) = \sum_{i=1}^{n}\left[\frac{\partial H}{\partial x_i}\frac{\partial H}{\partial p_i} + \frac{\partial H}{\partial p_i}\left(-\frac{\partial H}{\partial x_i} - k_{h_i}\right)\right] = -\sum_{i=1}^{n}\frac{\partial H}{\partial p_i}k_{h_i} \tag{53}$$

Equation (53) tell us that if L does not depend on t explicitly, generally H is not a first integral of generalized Euler equations but differs from a first integral by very small quantities $O(k_{hi})$.

In the general case we now evaluate for an arbitrary function $\Phi(x_1, x_2, ..., x_n, p_1, p_2, ..., p_n)$ the derivative

$$\frac{d\Phi}{dt} = \sum_{i=1}^{n}\left(\frac{\partial \Phi}{\partial x_i}\frac{dx_i}{dt} + \frac{\partial \Phi}{\partial p_i}\frac{dp_i}{dt}\right) = \sum_{i=1}^{n}\left[\frac{\partial \Phi_i}{\partial x_i}\frac{\partial H}{\partial p_i} - \frac{\partial \Phi}{\partial p_i}\frac{\partial H}{\partial x_i} - \frac{\partial \Phi}{\partial p_i}k_{h_i}\right] = \\ = [\Phi, H] - \sum_{i=1}^{n}\frac{\partial \Phi}{\partial p_i}k_{h_i} \tag{54}$$

where the expression

$$[\Phi, H] = \sum_{i=1}^{n}\left(\frac{\partial \Phi}{\partial x_i}\frac{\partial H}{\partial p_i} - \frac{\partial \Phi}{\partial p_i}\frac{\partial H}{\partial x_i}\right) \tag{55}$$

is the Poisson bracket of the functions Φ and H.

Equation (54) tell us that if [Φ,H]=0, generally Φ is not a first integral of generalized Euler equation but differs from an exact first integral by quantities $O(k_{hi})$.

We observe that the first integral which is approximately conserved, is the total energy (H=T+U) and the total angular momentum. Therefore the term $(\partial\Phi/\partial p_i)k_{hi}$ may represent the error in the evaluation of every motion constant. The error of the motion constant can be found in many applications involving electromagnetic problems. One case of application is the light propagation in a matter.

ELECTROMAGNETISM: AN EVALUATION FOR K_H AS RELATIVISTIC ERROR CONCERNING THE MEASURE OF THE LIGHT PROPAGATION IN A LIQUID

We first remember that the determination of the trajectory of a charged particle can be reduced to an optical problem with the introduction of a suitable refractive index, variable from point to point. Consider a particle of a charge *e* and of mass *m* moving in a steady electrostatic potential field ϕ(x,y,z). The laws of geometrical optics may be derived from Fermat's principle (Bergmann, 1942), according to which the path of light between two points P_1 and P_2 makes the optical length a minimum

$$\int_{P_1}^{P_2} n ds = \min . \tag{56}$$

The motion of a system of material points can be described in a variational form, by Hamilton's principle, which for the special case of a single material point is expressed by the condition (Bergmann, 1942)

$$\int_{P_1,t_1}^{P_2,t_2} L(\dot{x},\dot{y},\dot{z},x,y,z)dt = \min . \tag{57}$$

where L is the Lagrangian function.

For a relativistic electron, with charge *e* and mass *m* the general Lagrangian equation, according to the above analysis, may be expressed as

$$\frac{d\mathbf{p}}{dt} = e\left(\mathbf{E} + \frac{1}{c}\mathbf{v}\times\mathbf{B}\right) + \mathbf{k}_{\mathrm{h}} \tag{58}$$

which is the generalized Newton-Lorentz form of the equation of motion with the relativistic error k_h.

In this manner it is possible to evaluate the general electron-optical refractive index as (Landau, 1959)

$$n = \frac{mv_t}{\sqrt{1-\beta^2}} + \frac{e}{c}\mathbf{A}\cdot\mathbf{s} + k_h^{'} \tag{59}$$

where

$$k_h^{'} = \int_0^t k_h dt = k_h t \tag{60}$$

A·**s** is the component of the vector "potential" in the direction of motion, and $\beta = v_t / c$.

The (59) represents the equivalent refractive index of a liquid in which light propagates.

If we denote the speed of light by *c*, the velocity of the liquid by *v*, and the equivalent index of refraction by *n* of equation (59), we should expect that the observed speed of light is (Born et al., 1999)

$$u = \frac{c}{n} \pm v \tag{61}$$

the sign depending on the relative directions of the light and the flow. The actual result was that the change of speed of light is, within the limits of experimental error, given by (Born et al., 1999)

$$u = \frac{c}{n} \pm v\left(1 - \frac{1}{n^2}\right) \tag{62}$$

where the error in the measure is represented by $1/n^2$. By using the equivalent refractive index (59) it is possible to measure the velocity as

$$u = \frac{c}{\frac{mu}{\sqrt{1-\beta^2}} + \frac{e}{c}\mathbf{A}\cdot\mathbf{s} + k_h^{'}} \pm v\left(1 - \frac{1}{\left(\frac{mu}{\sqrt{1-\beta^2}} + \frac{e}{c}\mathbf{A}\cdot\mathbf{s} + k_h^{'}\right)^2}\right) \tag{63}$$

where $v_t = u$ represents the velocity in the liquid and *v* the velocity of the liquid, and $\beta = u/c$.

We observe that the error of the measured velocity

$$\frac{1}{n^2} = \frac{1}{\left(\frac{mu}{\sqrt{1-\beta^2}} + \frac{e}{c}\mathbf{A}\cdot\mathbf{s} + k_h^{'}\right)^2} \tag{64}$$

tends to infinity when it is satisfied the condition

$$k_h^{'} = k_h t = -\left(\frac{e}{c}\mathbf{A}\cdot\mathbf{s} + \frac{mu}{\sqrt{1-\beta^2}}\right) \tag{65}$$

In this manner it is possible to define the limit condition of the error $1/n^2$ by varying $k^{'}_{h}$, as shown in Figure 3, where for plotting we suppose

$$\frac{e}{c}\mathbf{A}\cdot\mathbf{s} + \frac{mu}{\sqrt{1-\beta^2}} = 1 \tag{66}$$

Figure 3. Behaviour of error $1/n^2$ by varying $k_h^{'}$

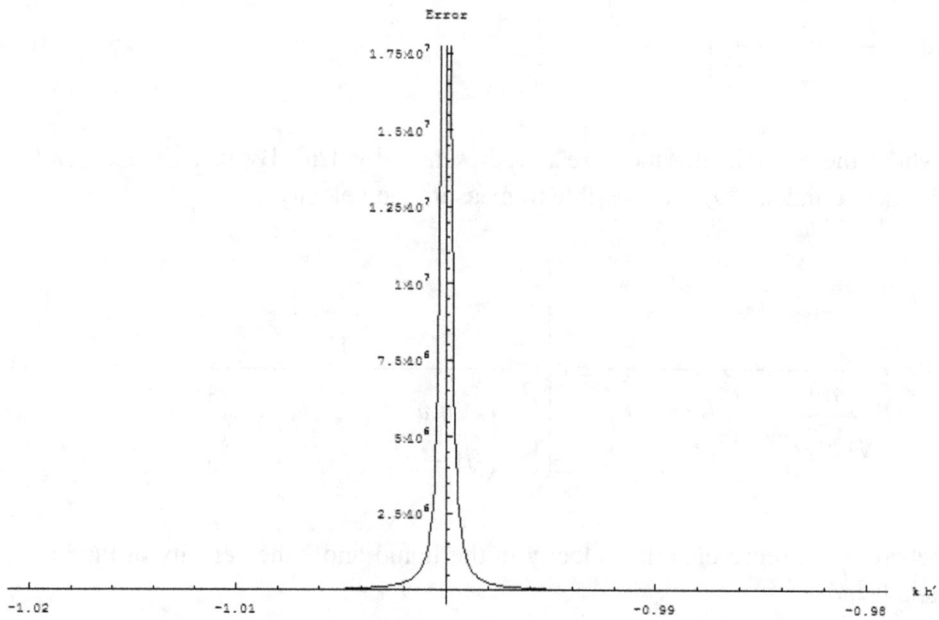

The result indicates that there is a particular measurement condition corresponding to the mathematical condition of the relativistic error, where light is not controllable. By considering that the refractive index is not easy to estimate in complex materials such as nanocomposite materials, and in general in unknown materials, we affirm that the wave propagation number could assume values related to equation (64). Different causes could affect the variation of the wave propagation number. Some of these are the topics listed in the following paragraph named "Future Research Direction and Discussions". The refractive index and consecutively the error of $1/n^2$ could change also in presence of uncontrollable physical phenomena involving for example nonlinear or impulsive variation of materials, environment behaviour (variation of temperature of the liquid where light propagates), coupling/interaction with other electromagnetic energies (optical, microwave or radio frequency wave). The study of the interaction between matter and electromagnetic wave is multifaceted and include many aspects which could represents perturbation of the light propagation system inducing the relativistic error. As example optical phenomena can be affected to light absorption, light scattering, multimode coupling, nonlinear coupling, enhancing by active materials, variation of quantum states and other. The full control of all perturbation causes is not possible and could happens that under particular condition and for particular time values the wavenumber satisfies the relativistic error condition.

FUTURE RESEARCH DIRECTIONS AND DISCUSSIONS

In this section we provide information about future research directions starting to relativistic errors. Some authors discussed that mathematical assumptions in quantum such as infinity assumption cannot prove uncertainty principle by any physical experiment (Subhendu, 2013). This aspect highlights that there are some conditions that cannot be proved by any measurements, but could represent uncertain physical phenomena or possible measurements errors. In other works some authors analyzed possible conceptualizations of measurement error and disturbance in quantum mechanics considering Heisenberg's error-disturbance relation (Bush et al., 2014). Starting to the same principle other researchers found a family of weighted uncertainty relations useful to solve incompatibility of observable quantum systems (Xiao et al., 2016). The uncertainty of the exact position and momentum of a particle during its motion, allow us to conclude that it is possible to stay in well-defined physical/mathematical condition representing error conditions and that the error can be due to an inexact solution of particle motion. The advanced research defined other extension of the concept of relativity of time: in particular the extend relativity (ER) represents the special relativity explaining the different acceleration due to the rate of a clock also in experiments (Friedman, 2013). Quantum uncertainty principle is also a recent topic: most existing entropic uncertainty relations use measures of joint uncertainty that yield themselves to a small class of operational interpretations (Narasimhachar et al., 2016). Recently, theoretical methods in quantum chemistry expanding the reach of the available relativistic methods to calculate all kinds of energy derivative properties have been applied (Autschbach, 2012). This enhances the broad range of the application field of relativistic effect which can be studied also in structural chemistry and spectroscopy. The anomalies in gravitational measurements involving electromagnetism have been discussed recently in a work concerning the variability of inertial mass (Nyambuya & Simango, 2014). In 2012 Howard E. M presented an overview of some open questions in general relativity with important consequences for causality theory is causal structure of the spacetime, focusing the attention on the problem of fundamental causal stability conditions (Howard, 2012). Yet in 1997 G. Svetlichny analyzed cases of light cone discontinuity in non-linear quantum mechanics, thus confirming the presence of non-relativistics limits in light propagation (Svetlichny, 1997). This aspect is similar to results found in this paper by confirming that the actual research needs of improvements. The researcher D. Sen discussed the existence of new 'measures' of uncertainty or indeterminacy in quantum measurement theory (Sen, 2014). As confirmed by J.P. Solovej, interaction of light and matter and in particular emission and absorption of light, are actual topics (Solovej, 2011). A recent book collected 21 papers about unsolved problems (involved relative motion) in special and general relativity, and in particular about the light principle conflicts with the notion that natural phenomena depend only upon mutual interaction (Yuhua et al., 2013). A chapter of this book discussed some light measurement mismatches about the refraction law which is not true in a rotating disk reference system. Actually Heisenberg uncertainty has been applied also in quantum information science for high-precision measurement and information security (Xiong et al., 2017).

CONCLUSION

The above results concerning the revised least action and generalized Euler equations represent: i) the possibility to analyze accurately the differences between some aspects between quantum mechanics and classical mechanics highlighting through the commutation parenthesis between the classical and quantum of observables

Table 1. Summary of research direction

Period	Reference	Topic
2017	(Xiong et al., 2017)	Heisenberg and measurements in quantum science (dingle trapped ion measurements)
2016	(Narasimhachar et al., 2016)	Uncertainty, joint uncertainty, and the quantum uncertainty principle
2016	(Xiao et al., 2016)	Heisenberg-Robertson and Schrödinger uncertainty relations
2014	• (Bush et al., 2014) • (Nyambuya & Simango, 2014) • (Sen, 2014)	• Heisenberg's principle analysis • Variability of inertial mass in measurements • Time, energy uncertainty relation
2013	• (Friedman, 2013) • (Yuhua et al., 2013)	• Extend relativity (ER) • Unsolved problems in general relativity
2012	• (Autschbach, 2012) • (Howard, 2012)	• Relativistic effect in chemistry and spectroscopy • Causal stability condition of spacetime
2011	(Solovej, 2011)	Stability of light/matter in quantum mechanics
1997	(Svetlichny, 1997)	Non-linear quantum mechanics

mechanics, ii) the possible violation of the symmetries in classical mechanics with the possibility to modify the first integral of the motion according with the proposed generalization of last action principle; iii) and also, as application, an accurate analysis of the error in the mechanic and electromagnetic measures. The topics discussed in this work are related to a possible explanation of relativistic errors concerning light measurement errors corresponding to a well-defined mathematical condition. The chapter proves the presence of a particular mathematical solution in classic mechanics uncertainty and proposes different topics of new research which could be correlated.

The theory found in this chapter could help to understand some measurements errors of light propagation supporting the knowledge of physical phenomena induced to materials, and unexpected motion errors in controlled systems such as particle accelerators.

ACKNOWLEDGMENT

In memory of my father Prof. Piero Adriano Massaro (Professor in Physics) who have collaborated in the development of the analytical approaches of this work.

REFERENCES

Autschbach, J. (2012). Perspective: Relativistic effect. *The Journal of Chemical Physics, 136*(1), 1–15. PMID:22519307

Bergmann, P. G. (1942). *Introduction to the Theory of Relativity*. Englewood Cliffs, NJ: Prentice-Hall, Inc.

Born, M., & Wolf, E. (1999). *Principle of Optics*. Cambridge, UK: Cambridge University Press, seventh expanded edition, Appendix II. doi:10.1017/CBO9781139644181

Bush, P., Lahti, P., & Werner, R. F. (2014). Colloquium: Quantum root-mean-square error and measurement uncertainty relations. *Reviews of Modern Physics, 86*(4), 1261–1281. doi:10.1103/RevModPhys.86.1261

Friedman, Y. (2013). Extending the relativity of time. Journal pf Physics. *Conference Series, 437*(1), 1–23.

Gelfand, I. M., & Fomin, S. V. (1963). *Calculus of Variations*. Englewood Cliffs, NJ: Prentice-Hall, Inc.

Goldstein, H. (1959). *Classical Mechanics*. Reading, MA: Addison Wesley Publishing Company, Inc.

Hawking, S. W. (1975). Particle creation by black holes. *Communications in Mathematical Physics, 43*(1), 199–220. doi:10.1007/BF02345020

Howard, E. M. (2012). Causal stability conditions for general relativistic spacetimes. *Journal of Physical Science and Application*, 2(7), 258–261.

Kaila, V. R. I., & Annila, A. (2008). Natural selection for least action. *Proceeding of The Royal Society A, 464*(1), 3055–3070. doi:10.1098/rspa.2008.0178

Landau, L. D., & Lifshits, E. M. (1959). *The Classical Theory of Fields*. Reading, MA: Addison Wesley Publishing Company, Inc.

Massaro, A. (2011). *Theory, Modeling, Technology and applications of Micro/Nano quantum electronic and photonic devices*. Kerala, India: Transworld Research Network.

Massaro, A., & Massaro, P. A. (2011). An observation on least action principle in classical mechanics oriented on the evaluation of relativistic error concerning the measurements of light propagating in a liquid. *International Journal of Measurement Technologies and Instrumentation Engineering, 1*(3), 14–24. doi:10.4018/ijmtie.2011070102

Narasimhachar, V., Poostindouz, A., & Gour, G. (2016). Uncertainty, joint uncertainty, and the quantum uncertainty principle. *New Journal of Physics*, *18*(1), 1–11.

Norton, J. D. (1993). General covariance and the foundations of general relativity: Eight decades of dispute. *Reports on Progress in Physics*, *56*(1), 791–858. doi:10.1088/0034-4885/56/7/001

Nyambuya, G. G., & Simango, W. (2014). On the gravitational bending of light. *International Journal of Astronomy and Astrophysics*, *4*(1), 250–263. doi:10.4236/ijaa.2014.41022

Oliveira, A. R. E. (2014). History of Two Fundamental Principles of Physics: Least Action and Conservation of Energy. *Advances in Historical Studies*, *3*(1), 83–92. doi:10.4236/ahs.2014.32008

Sen, D. (2014). The uncertainty relations in quantum mechanics. *Current Science*, *107*(2), 213–218.

Serway, R. A., Moses, C. J., & Moyer, C. A. (2005). *Modern Physics*. Belmont, WA: Thomson Learning, Inc.

Solovej, J. P. (2011). The stability of matter in quantum mechanics. *Bulletin of the American Mathematical Society*, *50*(1), 169–174. doi:10.1090/S0273-0979-2011-01366-0

Styer, D. F. (1995). Common misconceptions regarding quantum mechanics. *American Journal of Physics*, *64*(1), 31–34. doi:10.1119/1.18288

Subhendu, D. (2013). Assumptions in quantum mechanics. *International Journal and Mathematical Physics*, *3*(2), 53–68.

Svetlichny, G. (1997). On Relativistic Non-linear Quantum Mechanics. *Symmetry in Nonlinear Mathematical Physics*, *2*(1), 262–269.

Xiao, Y., J., Naihuan J., & Fei S. M. (2016). Weighted Uncertainty Relations. *Nature Scientific Reports* *6*(23201).

Xinzhong, W. (2016). Analytical mechanics structure of quantum mechanics and relativity space-time. *SCIREA Journal of Physics*, *1*(2), 108–124.

Xiong, T. P., Yan, L. L., Ma, Z. H., Zhou, F., Chen, L., Yang, W. L., ... Busch, P. (2017). Optimal joint measurements of complementary observables by a single trapped ion. *New Journal of Physics*, *19*(1), 1–9.

Yuhua, F., & Fengjuan, Z. (2013). *Unsolved Problems in Special and General Relativity*. Columbus, OH: Education Publishing.

KEY TERMS AND DEFINITIONS

Absorption (Electromagnetic Radiation): Way in which the energy of a photon is taken up by matter, typically the electrons of an atom.

Light Coupling: The lightwave is coupled with other electromagnetic signals.

Light Scattering: Form of scattering in which light in the form of propagating energy is scattered.

Particle Accelerator: Machine that uses electromagnetic fields to propel charged particles to nearly light speed and to contain them in well-defined beams.

Perturbation Theory: Mathematical methods for finding an approximate solution to a problem.

Uncertainty Principle: Any of a variety of mathematical inequalities asserting a fundamental limit to the precision with which certain pairs of physical properties of a particle, known as complementary variables, such as position x and momentum p, can be known.

Wavenumber: Spatial frequency of a wave, either in cycles per unit distance or radians per unit distance.

Chapter 3
A Step-Index Multimode Fiber-Optic Microbend Displacement Sensor Wavelength Dependent Loss

Sami D. Alaruri
Independent Researcher, USA

ABSTRACT

In this chapter, the wavelength dependence of bend loss in a step-index multimode optical fiber (100 μm core diameter; fused silica) was investigated for fiber bend radii ranging between 2.0 and 4.5 mm using six laser excitation wavelengths, namely, 337.1, 470, 590, 632.8, 750, and 810 nm. The results obtained from fitting the bend loss measurements to Kao's model and utilizing MATLAB® indicate that bend loss is wavelength dependent and transmission loss in multimode optical fibers increases with the decrease in the fiber bend radius. Furthermore, the response of a microbend fiber-optic displacement sensor was characterized at 337.1, 470, 632.8, 750, and 810 nm. Measurements obtained from the microbend sensor indicate that the sensor output power is linear with the applied displacement and the sensor output is wavelength dependent. Lastly, references for industrial and biomedical applications of microbend fiber-optic sensors are provided. Finally, a brief description for the transmission loss mechanisms in optical fibers is given.

DOI: 10.4018/978-1-5225-5406-6.ch003

INTRODUCTION

Loss introduced by bends in optical fibers has been recognized as a factor which contributes to the attenuation of the optical signal transmitted along an optical fiber (Marcuse, 1976a, 1976b; Berthold, 1995; Lagakos, Cole & Bucaro, 1987; Arya et al., 1995; Al-Aruri, 1991; Boechat, Hall & Jones, 1991; Donlogic, 2009; Schermer & Cole, 2007; Kao, 1982; Snyder, White & Mitchell, 1975; Lagakos & Bucaro, 1987). Decreasing the optical fiber bend radius increases the transmitted signal loss due to the transfer of power from the guided modes to the radiation modes supported by the core of the optical fiber. Considerable theoretical and experimental research has been conducted on the subject of wavelength dependence of bend loss in single-mode optical fibers (Marcuse, 1976a, 1976b; Berthold, 1995; Lagakos, Cole & Bucaro, 1987; Arya et al., 1995; Al-Aruri, 1991). However, due to the complex nature of bend loss in multimode fibers there are few publications in this area (Boechat, Hall & Jones, 1991; Donlogic, 2009; Schermer & Cole, 2007; Kao, 1982; Snyder, White & Mitchell, 1975). The mathematical models proposed by Kao (1982) and Snyder, White, and Mitchell (1975) are two examples of theoretical models dealing with the subject matter of bend loss in multimode optical fibers.

The primary goal of this work is to examine the wavelength dependence of bend loss in multimode optical fibers at low power levels (<10 mW) and to compare experimental measurements to the values predicted by Kao's model. Here, it is worth noting that Kao's model was selected in this work for modeling the bend loss in multimode optical fibers as a function of wavelength because of the model simplicity.

Additionally, this work provides a discussion for the results obtained from a displacement multimode optical fiber microbend sensor at 337.1, 470, 632.8, 750 and 810 nm laser wavelengths.

Microbend fiber-optic sensors utilize the modulation of the laser beam intensity guided through the optical fiber core as the transduction mechanism for detecting environmental parameters (Lagakos & Bucaro, 1987; Yao & Asawa, 1983; Donlagic & Zavrsnik, 1997). Owing to the several advantages of fiber-optic sensors (i.e., low cost, small weight and size, EMI/RF noise immunity) when compared with electric/electronic sensors, fiber-optic microbend-based sensors found their unique place into numerous industrial and biomedical applications (Matias, Ikezawa & Corre, 2017; Udd & Spillman, 2011; Kron, MacDoughall & Mendez, 2014). Some of these applications include variable optical attenuators (VOA) (DeBoynton & Uschitsky, 2001; Ao & Furuichi, 2006), which are widely used in fiber amplifiers and lasers (Alaruri, 2012), monitoring structures deformation (Ribiero et al., 2011), pressure (Roriz et al., 2013; Fields et al., 1980), temperature (Li et al., 2009; Berthold et al., 1987; Lumholt et al, 1991), stress (Remouche, Georges & Meyrueis, 2013), displacement (Alaruri, 2017; Lagakos et al., 1981), acoustics (such as hydrophones) and vibrations (Fields & Cole, 1980; Lagakos et al., 1982), strain (Weiss, 1989) and acceleration (Freal et al., 1987) measurements. In addition, microbend fiber-optic sensors are used in heartbeat and breathing rate monitoring (Chen et al., 2014; Yang et al., 2015) and in vehicle weight-in-motion detection (Bin & Xinguo, 2010).

Other transduction mechanism utilized in the sensing operation of fiber-optic sensors include: phase modulation, wavelength modulation, polarization, scattering, blackbody emission for temperature detection, fluorescence decay lifetimes, absorbance, and optical time domain reflectometry (OTDR) for detecting hydrocarbons and chemicals (Buerck & Sensfelder, 1999). A detailed description for fiber-optic sensors transduction mechanisms is provided in references (Matias, Ikezawa & Corres, 2017; Udd & Spillman, 2014; Spillman & Udd, 2014).

THEORETICAL BACKGROUND

In his model Kao (1982) argues that the overall loss in step-index multimode fibers due to bend loss is inversely proportional to the square of the optical fiber bend radius and can be expressed empirically as

$$\text{Loss} = A/R^2 \tag{1}$$

where A is an arbitrary constant and loss is the ratio of the sensor output power to the senor input power.

Further, the V number is a dimensionless parameter used for determining the number of guided modes support by the core of a step-index multimode fiber and can be expressed using the formula (Paschotta, 2010; Hecht, 1999; Cherin, 1982):

$$V = \frac{2\pi}{\lambda}\partial NA = \frac{2\pi}{\lambda}\partial\sqrt{n_{core}{}^{2} - n_{cladding}{}^{2}} \tag{2}$$

where a is the radius of the fiber core, NA is the fiber numerical aperture, λ is the wavelength transmitted along the optical fiber, n_{core} is the refractive index of the fiber core and $n_{cladding}$ is the refractive index of the cladding around the fiber core. It can be shown that for a single-mode fiber propagation V~2.405.

In addition, the deformer spacing (deformer plate tooth spacing), L, is expressed as (Spillman & Udd, 2014):

$$L = 2\pi\, a\, n/NA \tag{3}$$

where a is the fiber core diameter, n is the refractive index of the fiber core and NA is the fiber numerical aperture.

Typically, multimode optical fibers have a large core diameter (> 9 µm) and they support more than one guided mode per polarization direction. Whereas, single-mode optical fibers support only one guided mode per polarization direction and the core diameter for a single-mode fiber is typically <= 9 µm (Paschotta, 2010; Hecht, 1999; Cherin, 1982).

Additionally, transmission loss mechanisms in optical fibers can be broadly divided into three mechanisms: 1) absorption loss which can be caused by impurities in the fused silica such as hydroxyl (OH^-) and metal ions, 2) scattering which can be attributed to linear scattering (such as Rayleigh or Mie scattering) and non-linear scattering (such as stimulated Brillouin or Raman scattering), and 3) bend loss which can result from microbends or macrobends along the optical fiber (Al-Aruri, 1991; Paschotta, 2010; Hecht, 1999; Cherin, 1982).

MEASUREMENT SYSTEM

The experimental measurement system used for measuring bend loss in a step-index multimode optical fiber is depicted schematically in Figure 1. The measurement system consists of four major parts: CW chopped or pulsed laser sources, mode scrambler, mandrel for generating bends of known diameters along a section of the fiber and a photodiode detection system.

The laser sources used in this study are tabulated in Table 1. The laser beams emitted from the CW lasers were chopped and focused down into the input end of a 50 meters step-index multimode fiber (100µm core diameter/140 µm clad diameter; Radiant Communication Corp, p/n SFS-100; NA=0.22; fused silica) by the means of a microscope objective lens (Newport, 20X, NA =0.40). Here, it is worth mentioning that when pulsed lasers were used in the test setup the chopper was removed from the optical path. During the test the focused laser beams over-filled the input end of the multimode optical fiber. As shown in Figure 1, a fiber-optic mode scrambler (Newport, p/n FM1) was inserted into the optical path in order to establish equilibrium mode distribution (EMD) independent of the light launch condition into the optical fiber. The mode scrambler operates by introducing a series of micro-bends in a small section of the optical fiber causing a steady-state power distribution between the modes propagating along the fiber length. In the established power distribution between the fiber modes, the ratio of power carried in each mode is altered relative to a reference mode travelling along the fiber. Consequently, the ratio of power

Figure 1. Schematic showing the multimode fiber bend loss measurement system

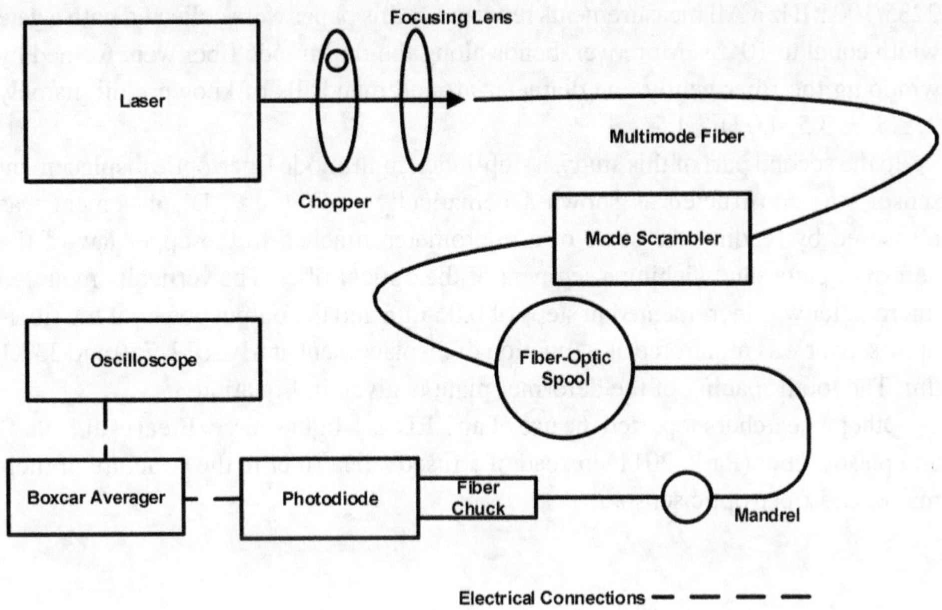

Table 1. List of the chopped CW and pulsed lasers used in this study

Wavelength (nm)	Power (mW)	Mode of Operation
810	4	Chopped at 71.4 Hz/Pulse width 7ms
750	4	Chopped at 71.4 Hz/Pulse width 7ms
632.8	10	Chopped at 71.4 Hz/Pulse width 7ms
590 (Rhodamine 590)	0.5	Pulsed/20 Hz (3 nsec FWHM)
470 (Coumarine 460)	0.5	Pulsed/20 Hz (3 nsec FWHM)
337.1	3	Pulsed/20 Hz (3 nsec FWHM)

distribution will no longer be dependent on the optical fiber length coordinates and a steady-state power distribution with equal amounts of power between the fiber modes can be achieved in the fiber core.

A PIN photodiode (Motorola, MRD 510) was used in the test setup to detect light emerging from the output end of the fiber. The rise-time of the PIN photodiode is less than 1 ns. The output of the PIN photodiode was fed into a boxcar averager (Princeton Applied Research Corporation, Model CW-1). The averaged output analog signal generated by the boxcar averager was read using an oscilloscope (Tektronix 2235/100 MHz). All measurements reported in this paper were collected with a gate width equal to 10 ms. Moreover, bends along the multimode fiber were formed by wrapping the fiber gently one diameter around mandrills of known radii, namely 2, 2.5, 3, 3.5, 4.0 and 4.5.

In the second part of this study, a step-index multimode fiber-optic displacement sensor was constructed as shown schematically in Figure 2. Displacement was measured by reading the scale of a micrometer attached to the upper jaw of the deformer plate sandwiching a segment of the optical fiber. The vertically mounted micrometer was incremented in steps of 0.05 mm and the output power of the fiber-optic sensor was monitored as a function of displacement at 810, 632, 750 and 337.1 nm. The tooth spacing of the deformer plate is given in Equation 3.

Other researchers reported the use of an LED as a light source (Freal et al., 1987) and plastic fiber (Park, 2011) instead of a fused silica fiber in the structure of their microbend fiber-optic sensor.

Figure 2. Block diagram showing the bend loss multimode fiber-optic displacement sensor setup

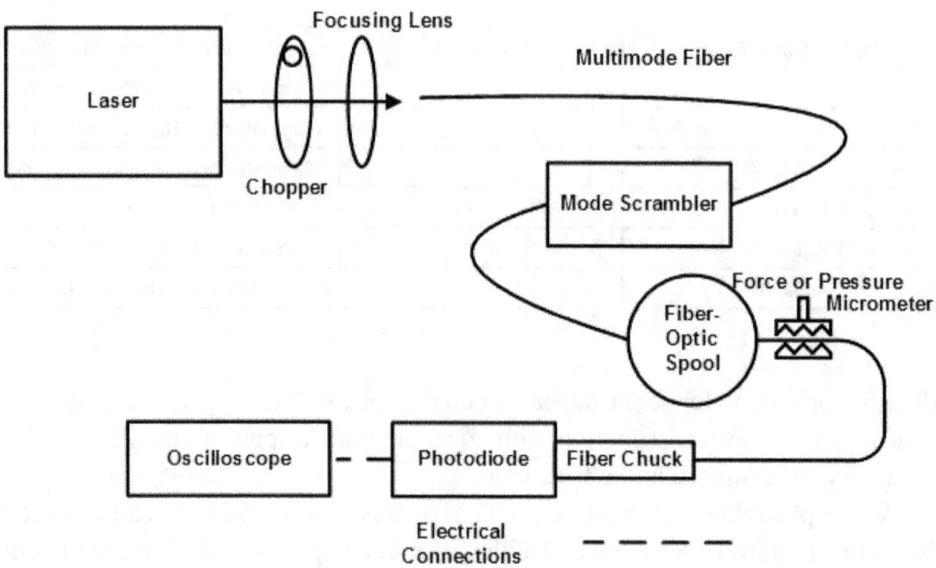

RESULTS AND DISCUSSION

As depict in Figure 3, the optical fiber transmission loss at 810, 750, 632.8, 590, 470 and 337.1 nm increased with the decrease in the multimode optical fiber bend radius. Further, the maximum and the minimum transmission loss values at 2 mm bend radius were 12% and 6% at 470 nm and 750 nm, respectively. The calculated constants for the tested wavelengths using the least-squares "lsqcurvefit" data fitting function provided in MATLAB ® (2015) are given in Table 2. The MATLAB ® script used for calculating the constants is listed in Table 3. The maximum and the minimum calculated coefficients were 44.037 and 22.03 nm for the 337.1 and 750 nm wavelengths, respectively. Generally, the calculated coefficients indicate that bend loss in step-index multimode fibers is wavelength dependent and the calculated coefficients have higher values at shorter wavelengths.

Figures 4 and 5 depict the normalized output power measurements obtained from the microbend fiber-optic sensor shown schematically in Figure 2 as a function of displacement at 810, 750, 632.8 and 337.1 nm. The obtained measurements indicate that the output of the sensor is linear with displacement. Further, the slopes of the calculated linear fits indicate they are wavelength-dependent. However, no clear correlation was found between the slopes and wavelength.

Figure 3. Plot showing the multimode fiber bend loss measurements collected at 810, 750, 632.8, 590, 470 and 337.1 nm. The solid lines are the empirical fits obtained for each set of data

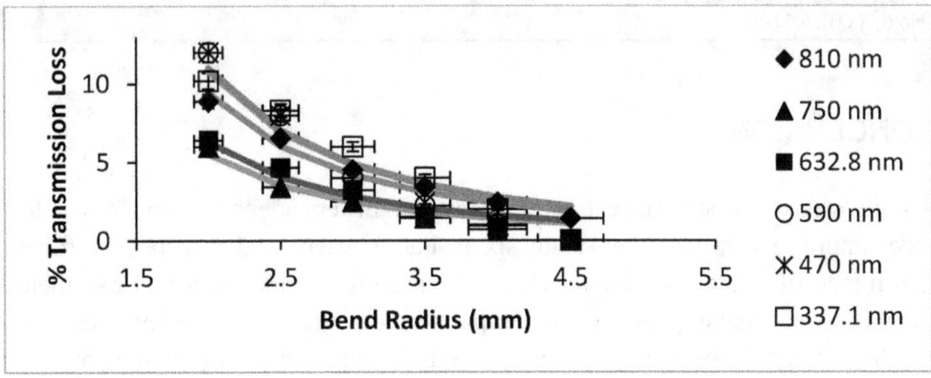

Table 2. The coefficients calculated for the experimental measurement at 810, 750, 632.8, 590, 470 and 337.1 nm employing Kao's (1982) model and the calculated V numbers for each wavelength

Wavelength (nm)	A	V
810	37.705	85
750	22.03	92
632.8	25.25	109
590	43.77	117
470	43.23	147
337.1	44.037	205

Table 3. MATLAB® code used for calculating the constants listed in Table 2

```
% read data
data=...
[2.0 8.9 2.5 6.55 3.0 4.5 3.5 3.5 4.0 2.4 4.5 1.4];
t = data(:,1); y = data(:,2);
% plot data
plot(t,y,'ro'); title('Data points')
% hold off
% define fit function F
F = @(x,xdata)(x(1)./xdata.^2);
% calculate coefficient
x0 =[0]; [x,resnorm,~,exitflag,output] = lsqcurvefit(F,x0,t,y);
disp('coefficients='); g=sprintf('%d\n',x);
fprintf('%s \n',g)
hold on
% plot the calculated function
plot(t,F(x,t)); hold off
```

CONCLUSION

A study was performed to investigate the wavelength dependence of bend loss in step-index multimode optical fibers using six excitation wavelengths. Also, a microbend multimode fiber-optic sensor which can be used for sensing physical parameters such as displacement, pressure and temperature has been developed and tested.

Based upon the obtained test results the following conclusions can be made:

1. Transmission loss in a step-index multimode fiber increases with the decrease in the fiber bend radius.

Figure 4. Plot depicting the collected measurements with the microbend fiber-optic displacement sensor at 810 nm and 750 nm. The solid lines are the empirical fits generated for each set of measurements

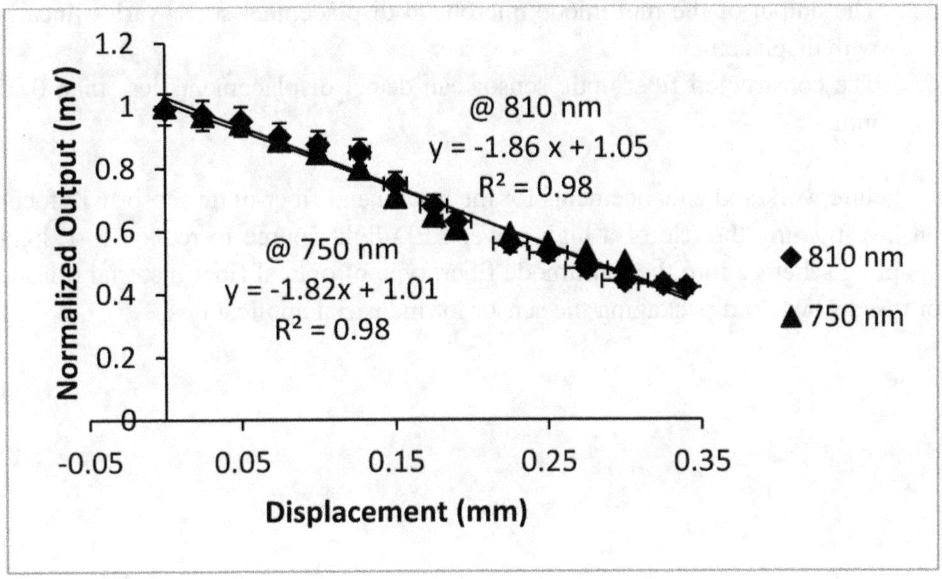

Figure 5. Plot illustrating the collected measurements with the microbend fiber-optic displacement sensor at 632.8 nm and 337.1 nm. The solid lines are the empirical fits generated for each set of measurements

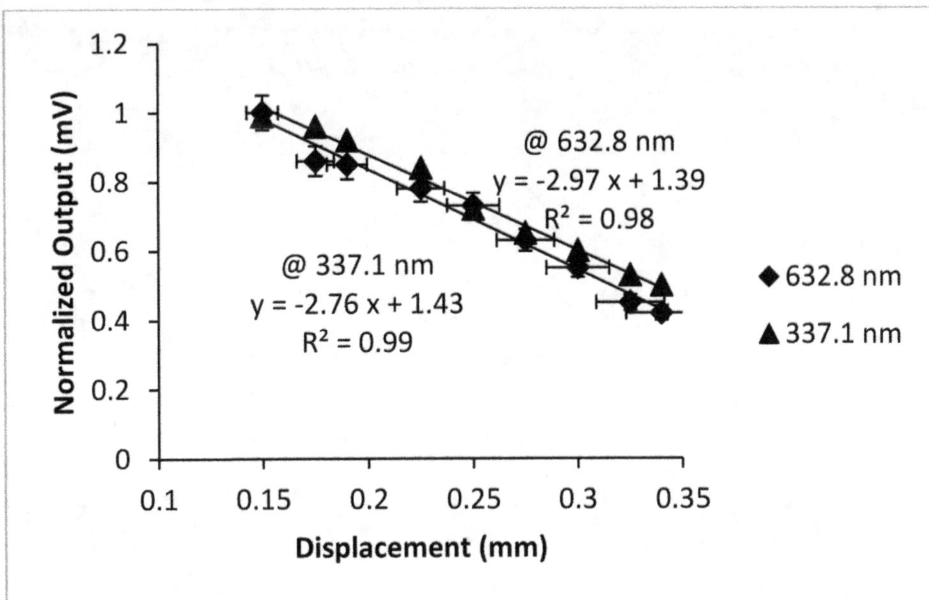

2. The Kao's model calculated coefficients indicate that bend loss in multimode optical fibers is wavelength dependent and the values of the coefficients are higher at shorter wavelengths.
3. The output of the multimode microbend displacement senor varies linearly with displacement.
4. The constructed fiber-optic sensor can detect displacements less than 0.05 mm.

Future work and enhancements for the microbend fiber-optic sensor will focus on investigating the use of a high-power LED light source to reduce cost, light coupling schemes into the multimode fiber, type of optical fiber material (plastic or fused silica) and packaging the sensor for industrial applications.

REFERENCES

Al-Aruri, S. D. (1991). *Studies of wavelength-dependent loss effects in optical fiber components and sensors* (MS thesis). Rose-Hulman Institute of Technology. (Report# CASOS-TR-44)

Alaruri, S. D. (2012). Gain, noise figure and efficiency characteristics of an L-band erbium doped fiber amplifier. *International Journal of Measurement Technologies and Instrumentation Engineering*, *2*(4), 1–11. doi:10.4018/ijmtie.2012100101

Alaruri, S. D. (2017). Study of wavelength-dependent bend loss in a step-index multimode fiber-optic microbend displacement sensor. *International Journal of Measurement Technologies and Instrumentation Engineering*, *6*(1), 13–21. doi:10.4018/IJMTIE.2017010102

Ao, M., & Furuichi, S. (2006). *Variable optical attenuator*. US Patent No. 7,065,284 B2.

Arya, V., Murphy, K. A., Wang, A., & Claus, R. O. (1995). Microbend losses in single mode optical fibers: Theoretical and experimental investigation. *Journal of Lightwave Technology*, *13*(10), 1998–2002. doi:10.1109/50.469736

Berthold, J. W. (1995). Historical review of microbend fiber-optic sensors. *Journal of Lightwave Technology*, *13*(7), 1193–1199. doi:10.1109/50.400697

Berthold, J. W., Ghering, W., & Varshneya, D. (1987). Design and characterization of a high temperature fiber-optic pressure transducer. *Journal of Lightwave Technology*, *LT-5*(7), 870–876. doi:10.1109/JLT.1987.1075594

Bin, M., & Xinguo, Z. (2010). Study of vehicle weight-in-motion system based on fiber-optic microbend sensor. *IEEE International Conference on Intelligent Computation Technology and Automation*, 458-461. 10.1109/ICICTA.2010.631

Boechat, A. D., Hall, S. D., & Jones, J. (1991). Bend loss in large core multimode optical fiber beam delivery systems. *Applied Optics*, *30*(3), 321–327. doi:10.1364/AO.30.000321 PMID:20581985

Buerck, J., & Sensfelder, E. (1999). Optical fiber sensors for the distributed measurement of hydrocarbons. *Proc. SPIE*, 3540, 98-109. 10.1117/12.339785

Chen, Z., Lau, D., Teo, J. T., Ng, S. H., Yang, X., & Kei, P. L. (2014). Field test of microbend fiber sensor for hospital use. *Proc. SPIE 9157, 23rd International Conference on Optical Fibre Sensors*, 91579A.

Cherin, A. H. (1983). *An Introduction to Optical Fibers*. New York: McGraw-Hill Book Company.

DeBoynton, W. L., & Uschitsky, M. (2001). *Fiber coupler variable optical attenuator*. US Patent No. 6,173,106 B1.

Donlagic, D. (2009). A low bending loss multimode fiber transmission system. *Optics Express*, *17*(24), 22081–22095. doi:10.1364/OE.17.022081 PMID:19997454

Đonlagić, D., & Završnik, M. (1997). Fiber-optic microbend sensor structure. *Applied Optics*, *22*(11), 837–839. PMID:18185679

Fields, J. N., & Cole, J. H. (1980). Fiber microbend acoustic sensor. *Applied Optics*, *19*(3), 265–3267. PMID:20234597

Fields, N. J., Asawa, C. K., Ramer, O. G., & Barnoski, M. K. (1980). Fiber optic pressure sensor. *The Journal of the Acoustical Society of America*, *67*(3), 816–818. doi:10.1121/1.383957

Freal, J. B., Zarobila, C. J., & Davis, C. M. (1987). A microbend accelerometer for borehole deployment. *Journal of Lightwave Technology*, *LT-5*(7), 993–996. doi:10.1109/JLT.1987.1075609

Hecht, J. (1999). *Understanding Fiber Optics* (3rd ed.). London: Prentice-Hall International.

Kao, C. K. (1982). *Optical fiber systems: Technology*. New York: Design and Applications, McGraw-Hill Company.

Krohn, D., MacDougall, T., & Mendez, A. (2014). *Fiber optic sensors fundamentals and applications* (4th ed.). Bellingham, WA: SPIE Press. doi:10.1117/3.1002910

Lagakos, N., & Bucaro, J. A. (1987). Optimizing fiber optic microbend sensor. *Proceedings of the Society for Photo-Instrumentation Engineers*, *718*, 12–20. doi:10.1117/12.937494

Lagakos, N., Cole, J. H., & Bucaro, J. A. (1987). Microbend fiber-optic sensor. *Applied Optics*, *26*(11), 2171–2180. doi:10.1364/AO.26.002171 PMID:20489839

Lagakos, N., Litovitz, T., Macedo, P., Mohr, R., & Meister, R. (1981). Multimode optical fiber displacement sensor. *Applied Optics*, *20*(2), 167–168. doi:10.1364/AO.20.000167 PMID:20309083

Lagakos, N., Trott, W., Hickman, T., Cole, J., & Bucaro, J. (1989). Microbend fiber optic sensor as extended hydrophone. *IEEE Journal of Quantum Electronics*, *OF-18*, 1633–1638.

Li, X., Chena, Y., Dengb, Y., Lia, X., & Lia, J. (2009). Novel microbending optical fiber temperature sensor. *Proc. of SPIE*, 7508, 75080N-1 to 75080N-5. 10.1117/12.839634

Lumholt, O., Bjarklev, A., Dahl-Petersen, S., Larsen, C. C., Povlsen, J. H., Rasmussen, T., & Rottwitt, K. (1991). Simple fiber optic low-temperature sensor that uses micro-bending loss. *Optics Letters*, *16*(17), 1355–1357. doi:10.1364/OL.16.001355 PMID:19776969

Marcuse, D. (1976a). Curvature loss formula for optical fibers. *Journal of the Optical Society of America*, *66*(3), 216–220. doi:10.1364/JOSA.66.000216

Marcuse, D. (1976b). Field deformation and loss caused by curvature of optical fibers. *Journal of the Optical Society of America*, *66*(4), 311–320. doi:10.1364/JOSA.66.000311

Matias, I. R., Ikezawa, S., & Corres, J. (2017). Fiber Optic Sensors: current status and future possibilities. Springer.

MATLAB. (2015). *Primer, R2015b*. The Math Works, Inc. Retrieved from http://www.mathworks.com

Park, J. (2011). Plastic optical fiber sensor for measuring driver-gripping force. *Optical Engineering Letters, 50*(2), 02050-1 - 02050-3.

Paschotta, R. (2010). *Optical Fiber technology, SPIE Field Guides, Vol. FG16*. Bellingham, WA: SPIE Press. doi:10.1117/3.853722

Remouche, M., Georges, F., & Meyrueis, P. (2013). Stress sensing by an optical fiber sensor: Method and process for the characterization of the sensor response depending on several designs. *Optics and Photonics Journal*, *3*(02), 194–203. doi:10.4236/opj.2013.32032

Ribeiro, L. A., Rosolem, J. B., Dini, D. C., Floridia, C., Hortencio, C. A., da Costa, E. F., . . . Durelli, A. S. (2011). Fiber optic bending loss sensor for application on monitoring of embankment dams, Microwave & Optoelectronics Conference (IMOC), 2011 SBMO/IEEE MTT-S International, 637-641.

Roriz, P., Frazao, O., Lobo-Ribeiro, A.B., Santos, J.L. & Simoes, J.A. (2013). Review of fiber-optic pressure sensors for biomedical applications and biomechanical applications. *Journal of Biomedical Optics, 18*(5), 050903-1 to 050903-18.

Schermer, R. T., & Cole, J. H. (2007). Improved bend loss formula verified for optical fiber by simulation and experiment. *IEEE Journal of Quantum Electronics, 42*(10), 899-909.

Snyder, W., White, I., & Mitchell, D. J. (1975). Radiation from bent optical waveguides. *Electronics Letters, 11*(15), 332–333. doi:10.1049/el:19750254

Spillman, W. B., & Udd, E. (2014). Field guide to fiber optic sensors (vol. FG34). SPIE Press.

Udd, E., & Spillman, W. B. (2011). *Fiber optic sensors: an introduction for engineers and scientists* (2nd ed.). John Wiley & Sons, Inc. doi:10.1002/9781118014103

Weiss, J. D. (1989). Fiber optic strain gage. *Journal of Lightwave Technology, 7*(9), 1308–1318. doi:10.1109/50.50709

Yang, X., Chen, Z., Elvin, C. S. M., Janice, L. H. Y., Ng, S. H., Teo, J. T., & Wu, R. (2015). Textile fiber optic microbend sensor used for heartbeat and respiration monitoring. *IEEE Sensors Journal, 15*(2), 757–761. doi:10.1109/JSEN.2014.2353640

Yao, S. K., & Asawa, C. K. (1983). Microbending fiber optic sensing. *Proceedings of the Society for Photo-Instrumentation Engineers, 412*, 9–12. doi:10.1117/12.935788

Section 2

Measurement Strategies in Industrial Applications and Development of Suitable Reference Materials

This section describes the significance of measurement technology in industry through three model examples. Consequently, the need for appropriate reference materials and measurement methods is highlighted in the concluding chapters of this section.

Chapter 4
Analysis of Sidebands Failures in Asynchronous Drives

K. Vinoth Kumar
Karunya University, India

Prawin Angel Michael
Karunya University, India

ABSTRACT

This chapter deals with the diagnosis of induction motors (IM) with the so-called motor current signature analysis (MCSA). The MCSA is one of the most efficient techniques for the detection and the localization of electrical and mechanical failures, in which faults become apparent by harmonic components around the supply frequency. This chapter presents a summary of the most frequent faults and its consequences on the stator current spectrum of an IM. A three-phase IM model was used for simulation taking into account in one hand the normal healthy operation and in the other hand the broken rotor bars, the shorted turns in the stator windings, the voltage unbalance between phases of supply, and the abnormal behavior of load. The MCSA is used by many authors in literature for faults detection of IM. The major contribution of this work is to prove the efficiency of this diagnosis methodology to detect different faults simultaneously, in normal and abnormal functional conditions. The results illustrate good agreement between both simulated and experimental results.

DOI: 10.4018/978-1-5225-5406-6.ch004

INTRODUCTION

Induction motors play an important role in industry for the rotating machine practice because of their robustness, low costs and quasi-absence of maintenance. Nevertheless, it is seen that this machine presents an electric or mechanical defect. The faults of these machines are varied. However the most frequent are (Benbouzid, 2000; Razik, 2002; Trajin et al., 2008): opening or shorting of one or more of a stator phase winding, broken rotor bar or cracked rotor end rings, static or dynamic air-gap irregularities, and bearing failures. In order to avoid such problems, these faults have to be detected to prevent a major failure from occurring. It is well known that a motor failure may yield an unexpected interruption at the industrial plant, with consequences in costs, product quality, and safety. During the past twenty years, there has been a substantial amount of fundamental research into the creation of condition monitoring and diagnostic techniques for IM drives.

Different detection approaches are proposed in the literature such as those based on the Extended Park's Vector Approach (EPVA), which allows the detection of inter-turn short circuits in the stator winding (Acosta et al., 2004). The EPVA is appropriate for the stator winding monitoring. Çalis and Çakir (2007) used the 2.s.fs spectral component in the stator current zero crossing times (ZCT) spectrum as an index of rotor bar faults. However, the major deficiency for this fault indicator, for low slip IM Operating at no load condition is that it may then be difficult to read its value. In Casimir et al. (2006), the authors studied the diagnosis of IM by pattern recognition method. This method consists in extracting features from the combination of the stator currents and voltages. Some of these features could be irrelevant or redundant. Therefore, the Sequential Backward Selection (SBS) algorithm is applied to the complete set of features to select the most relevant. Then they used the k-Nearest Neighbours (kNN) rules to monitor the IM functioning states. This rule is applied with reject options in order to avoid automatic classifications and diagnosis errors. Didier et al. (2007) employed the Fourier Transform of the stator current and then analyzed its phase. It is shown that the calculated phase gives good results when the motor operates near its nominal load. For weak load, the results obtained are not robust enough for the detection of an incipient rotor fault.

In Li and Mechefske (2004), the authors used the vibration monitoring methodology to detect incipient failures in IM. Vibration monitoring system requires storing of a large amount of data. Vibration is often measured with multiple sensors mounted on different parts of the machine. The examination of data can be tedious and sensitive to errors. Also, fault related machine vibration is usually corrupted with structural machine vibration and noise from interfering machinery. To overcome these problems Poyhonen et al. (2003) used the Independent Component Analysis

(ICA) to compress measurements from several channels into a smaller amount of channel combinations and to provide a robust and reliable fault diagnostics routine for a cage IM. This paper is focused on the Motor Current Signature Analysis (MCSA) approach. This technique utilizes results of spectral analysis of the stator current (precisely, the supply current) of an IM to spot an existing or incipient failure of the motor or the drive system. It is claimed that MCSA monitoring is the most reliable method of assessing the overall health of a rotor system (Thomson, 2001). Unlike the other techniques, MCSA can provide the same indications without requiring access to the motor.

DIFFERENT CATEGORIES OF FAILURES

Broken Rotor Bars

It is well known that, under normal conditions of working, a 3-phase IM with symmetrical stator winding fed from a symmetrical supply voltage with frequency fs, will produce a resultant forward rotating magnetic field at synchronous speed and if exact symmetry exists there will be no resultant backward rotating field. When rotor defect appears, it creates in addition of the direct rotor field an inverse field that turns to the speed (-s.ωs). It is due to the fact that the rotor currents are now direct and inverse following the unbalance of resistances. It is the interaction of this field with the one descended of stator windings that induces an e.m.f. and current in the stator winding at (1-2.s).fs. This cyclic current variation causes a speed oscillation at twice the slip frequency (2.s.fs) and finally, this speed oscillation induces, in the stator current spectrum, an upper component at (1+2.s).fs, and so on (Çalis & Çakir, 2007; Mehala & Dahiya, 2009). Therefore, broken rotor bars induce harmonic components in the stator current at frequencies given by (Thomson, 2009; Jung et al., 2006):

$$f_{bb} = \left(1 \pm 2ks\right) f_s \tag{1}$$

where: $k = 1, 2, 3\ldots, K \in \mathbb{N}$,

$_{bb}$: broken rotor bar frequency,
f_s: electrical supply frequency,
p: number of pole pairs, s: slip.

Unbalanced Supply Voltage

Asymmetrical stator faults (caused by stator winding faults or asymmetrical supply voltages) are also common in IM (Messaoudi et al., 2007). An asymmetrical stator supply voltage can be caused by the opening of one of the three phases, by the presence of one-phase-load in the environment near of the motor, or by the source. The consequences of an unbalanced supply voltage applied to a three phase IM are the reduction of the useful torque and the increase of the losses. Unbalances result in an inverse component that generates high rotor current provoking a very important heating of the rotor and implying an overheating of the motor. The calculation of the unbalance can be approached by the following equation:

$$unbalance(\%) = 100 \times Max\left(\frac{V_h - V_{av}}{V_{av}}, \frac{V_{av} - V_l}{V_{av}}\right) \quad (2)$$

where:

V_h: highest voltage, and
V_l: lowest voltage,

An unbalanced in the supply voltage induces sidebands in stator current spectrum of the IM at the following frequencies:

$$f_{usv} = (1 \pm 2k) f_s \quad (4)$$

where: $k = 1, 2, 3\ldots, K \in \mathbb{N}$, f_{usv}: unbalanced supply voltage frequency, f_s: electrical supply frequency.

Fluctuation in Load Torque

The load torque variation induces components in the current spectrum which coincide with those caused by a fault condition. In an ideal machine where the stator flux linkage is purely sinusoidal, any oscillation in the load torque at a multiple of the rotational speed will produce stator currents at frequencies of (Benbouzid, 2000; Fenger et al., 2003):

$$f_{lo} = f_s \pm kf_r = \left[1 \pm k\left(\frac{1-s}{p}\right)\right] f_1 \tag{5}$$

where: $k = 1, 2, 3\ldots, K \in \mathbb{N}, f_{bb}$

f_{lo}: load effects frequency,
f_s: electrical supply frequency,
f_r: mechanical rotor speed in Hertz,
p: number of pole pairs, s: slip.

Shorted Turns in Stator

Asymmetrical Inter-turn short circuits in stator windings constitute a category of faults that is most common in induction motors. Typically, short circuits in stator windings occur between turns of one phase, or between turns of two phases, or between turns of all phases. Moreover, short circuits between winding conductors and the stator core also occur. Studies (Thomson, 2001; Blodt et al., 2006) prove that the stator current is enriched by short turns. The additional components are at the following frequencies:

$$f_{st} = \left[\frac{n}{p}\left(1-s\right) \pm k\right] f_s \tag{6}$$

where: $k = 1, 2, 3\ldots, K \in \mathbb{N}$,

f_{st}: short turns frequency,
f_s: electrical supply frequency,
p: number of pole pairs, s: slip.

SIMULATION AND EXPERIMENTAL RESULTS

Motor Test Ring

The Experimental setup shown in Figure 1 is also instrumented with Hall-effect current transducers. The motor current spectrum analysis algorithm is implemented

Figure 1. Experimental setup

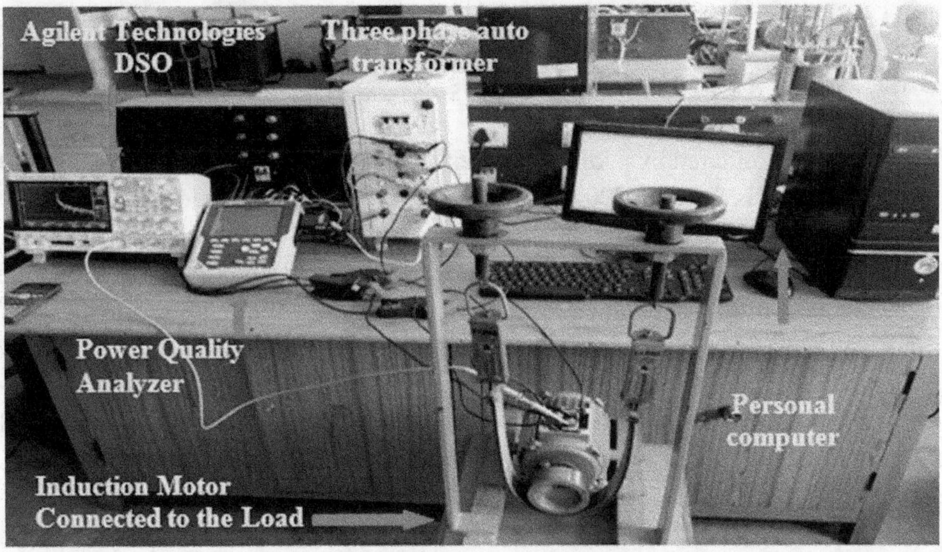

using MATLAB software package. The acquisition of stator current takes place at a sampling frequency of 10 kHz on a length of 10 seconds. The experimental setup consists of two IMs; the first one is a 0.375 kW, 4-poles, 22-rotor bars IM. Figure 2 shows how damage in one of the rotor bars was physically seeded. It was performed by drilling through one of the aluminum conductors that made up the squirrel cage. The second motor used is a 3 kW IM, in which, two extra outputs are created as shown in Figure 2. These added outputs represent respectively, the extremities of a

Figure 2. Seeded broken rotor bar fault and seeded short-circuited windings in one stator phase

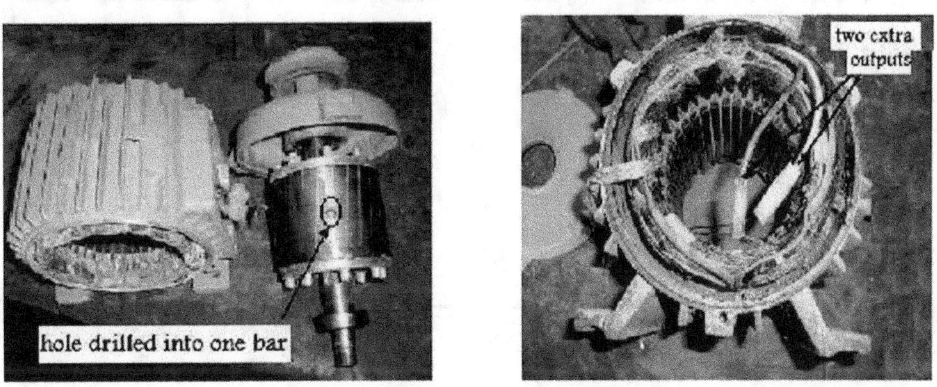

stator phase with 5 and 8 short-circuited-windings. The load is a DC motor acted as a generator and its power output was dissipated in a variable resistor bank.

Case Study 1: Broken Rotor Bars

Rotor fault is simulated by adding extra resistance to the single phase resistance of rotor circuit (Çalis & Çakir, 2007; Didier & Razik, 2001; Treetrong, 2010). In Figure 4, spectrum of stator current under 75% operational load is given, respectively, for healthy motor, rotor resistance increased 20% of its rated value, and rotor resistance increased 40% of its rated value. Referring to the healthy motor spectrum given by Figure 3 (a), we can clearly see the sidebands around the fundamental frequency in Figure 3 (b) and (c). These sidebands are the results of the defect created in the rotor. Frequencies of these sidebands correspond precisely to the mathematical relation $(1\pm2.s).f_s$ given by (1). The amplitude and the number of the sidebands frequencies components are proportional to the amount of broken rotor bars. Table 1, shows the harmonic components of the stator current under 75% of rated load with two levels of rotor asymmetry.

When the IM operates with no load as shown in Figure 4 (a), the algorithm of detection does not give any response. This means that the jump located at the frequency $(1\pm2.k.s).f_s$ is not detectable due to its low amplitude. The current which crosses the rotor bars is not important enough to create a consequent jump at this frequency.

For an IM with a constant number of broken rotor bars operated under different levels of operational load, it is clearly seen in Figure 4 that the amplitudes of the

Figure 3. Stator current spectrum under 75% operational load: (a) healthy motor, (b) $R_{ra} = R_r + 20\% R_r$, (c) $R_{ra} = R_r + 40\% R_r$

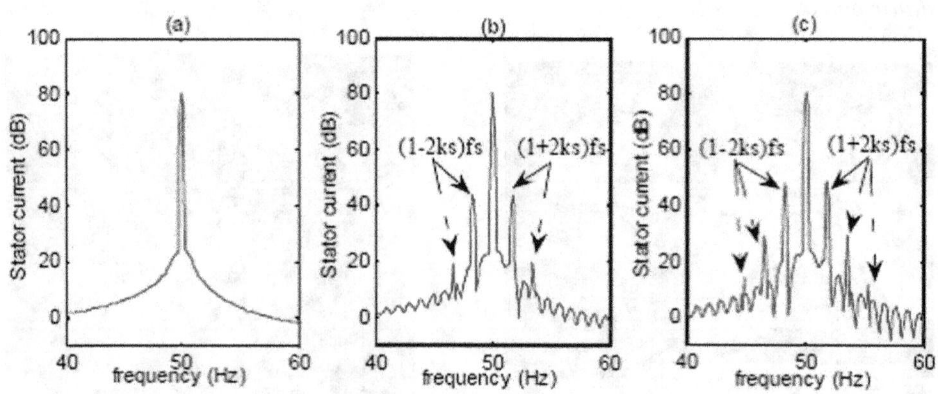

Table 1. Harmonic components of the stator current

Frequencies	$R_{ra} = R_r + 20\% R_r$	$R_{ra} = R_r + 40\% R_r$
$(1 - 6.s).f_s$		44,75 Hz
$(1 - 4.s).f_s$	46,70 Hz	46,50 Hz
$(1 - 2.s).f_s$	48,35 Hz	48,25 Hz
$(1 + 2.s).f_s$	51,65 Hz	51,75 Hz
$(1 + 4.s).f_s$	53,30 Hz	53,50 Hz
$(1 + 6.s).f_s$		55,25 Hz

Figure 4. Stator current spectrum with $R_{ra} = R_r + 40\% R_r$: (a) with zero load condition, (b) with 25% load condition

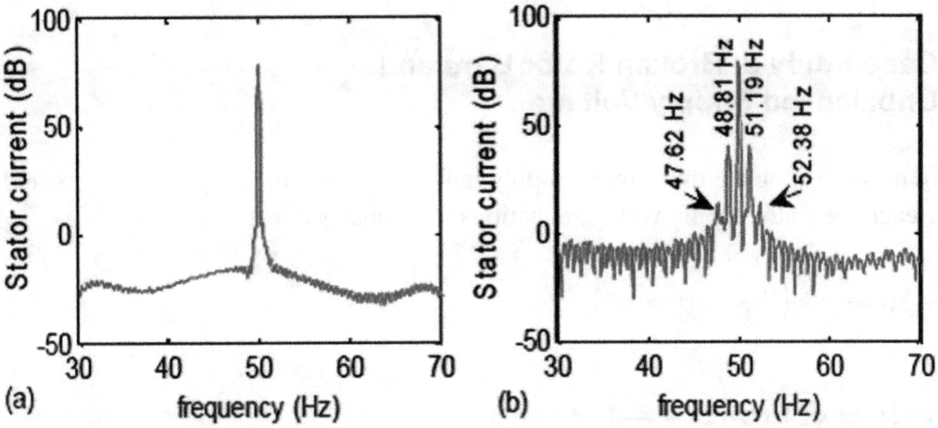

sidebands around the supply frequency increase as the load torque rises. Thus, this indicator component is sensitive to the load variation. Therefore, it is not exactly reflecting the degree of fault. But, the amplitude of the sidebands can be used as an efficient indicator of faults degree if the load effect is taken into account. The frequencies of the first two sidebands around the fundamental frequency for an IM operating with a fixed number of broken rotor bars under 25% of its rated load are 48.81 Hz and 51.19 Hz. Figure 6 illustrates the stator current spectrum when one broken rotor bar fault was experimentally seeded. When, the motor is under 25% of its rated load as shown in Figure 5 (a), the amplitude jump at frequencies $(1 \pm 2.k.s).f_s$ due to rotor fault appeared in the stator current spectrum. Despite, the fault existence (one broken rotor bar), the MCSA failed to give a positive result in the case of unloaded motor Figure 5 (b).

Figure 5. Experimental stator current spectrum of an IM with broken rotor bar: (a) under 25% of load, (b) unloaded motor

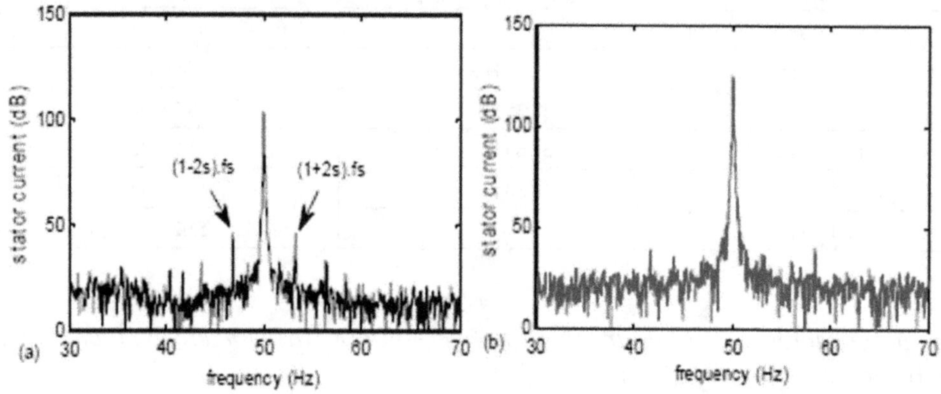

Case Study 2: Broken Rotor Bars and Unbalanced Supply Voltage

In order to simulate unbalanced supply faults, a V_0 drop in one phase was created. Hence the stator supply voltage equations were changed as follows:

$$V_{as}\left(t\right) = \sqrt{2}\left(V - V_o\right)\cos\left(\omega t\right),$$

$$V_{bs}\left(t\right) = \sqrt{2}\,V\cos\left(\omega t - \frac{2\pi}{3}\right),$$

$$V_{cs}\left(t\right) = \sqrt{2}\,V\cos\left(\omega t + \frac{2\pi}{3}\right) \tag{7}$$

In Figure 6 (a), new components at frequencies $(1+2k).f_s$: (150 Hz; 250 Hz) emerge in the current spectrum. These harmonic components are induced by the unbalance in the supply voltage. Figure 6 (b) and (c), clearly, show that symptoms of both unbalanced stator voltage supply $(1+2.k).f_s$ and broken rotor bars $(1\pm2.k.s).f_s$ appear in the stator current spectrum.

Figure 7 (a) and (b) illustrate the experimental results of the stator current spectrum when the IM is operated under different levels of unbalanced supply voltage. It

Figure 6. Stator current spectrum for 20V drop of one phase voltage: (a) healthy rotor; (b) broken rotor bars ($R_{ar} = R_r + 20\% R_r$); (c) broken rotor bars ($R_{ar} = R_r + 40\% R_r$)

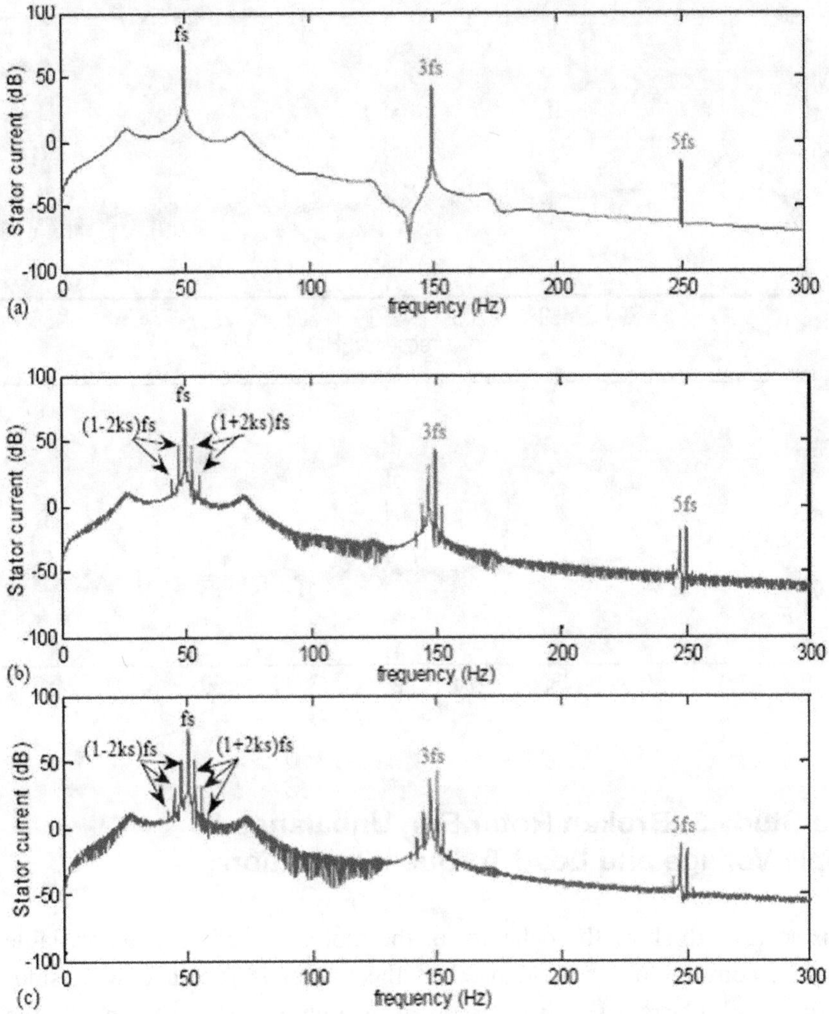

appears in Figure 7 (a) and Figure 7 (b) in the sidebands at frequencies (1+2*k*).*fs*. It is clearly seen that the magnitudes of these sidebands rise when the drop in the supply voltage is increased. The rise in the magnitude of sidebands created by the unbalanced supply voltage is clearer at the supply frequency third harmonic (150 Hz).

Figure 7. Experimental stator current spectrum of IM operating under unbalanced supply voltage: (a) partial unbalance (20% drop in one supply voltage phase); (b) full unbalance (opening of one phase)

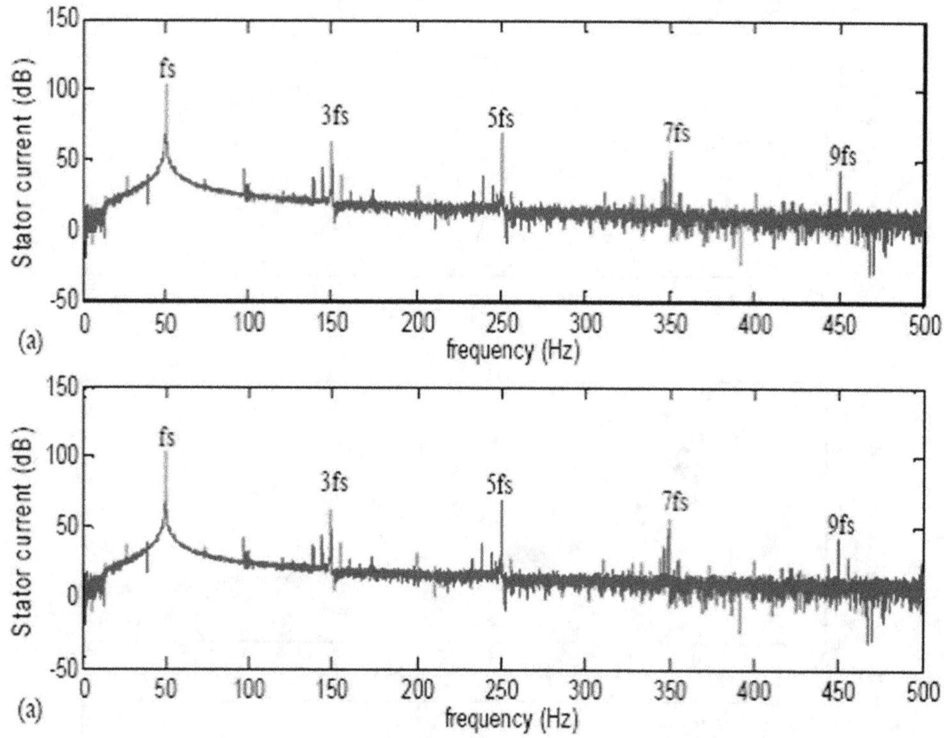

Case Study 3: Broken Rotor Bar, Unbalanced Supply Voltage and Load Torque Fluctuation

In this section, to show the behavior of the motor operating under variable load condition, constant motor load used in the model is replaced with sinusoidal changing load. Current spectrum is given in Figure 7. It is shown that the load variation produces sidebands around the supply frequency equal to frequencies of ($f_s \pm k.f_r$). Figure 8 depicts the current spectrum of an IM operating under three different operational conditions. In Figure 8 (a), only the frequencies due to load variations appeared in the stator current spectrum. In Figure 8 (b), in addition of the frequencies components due to load variations, sidebands also appeared around the fundamental at frequencies $(1 \pm 2.k.s).f_s$ produced by the rotor asymmetry. In Figure

Figure 8. Stator current spectrum for motor operating under variable load condition: (a) healthy motor; (b) motor with broken rotor bars ($R_{ar} = R_r + 40\% R_r$); (c) motor with broken rotor bars ($R_{ar} = R_r + 40\% R_r$) and 44 V drop in one phase of the supply voltage

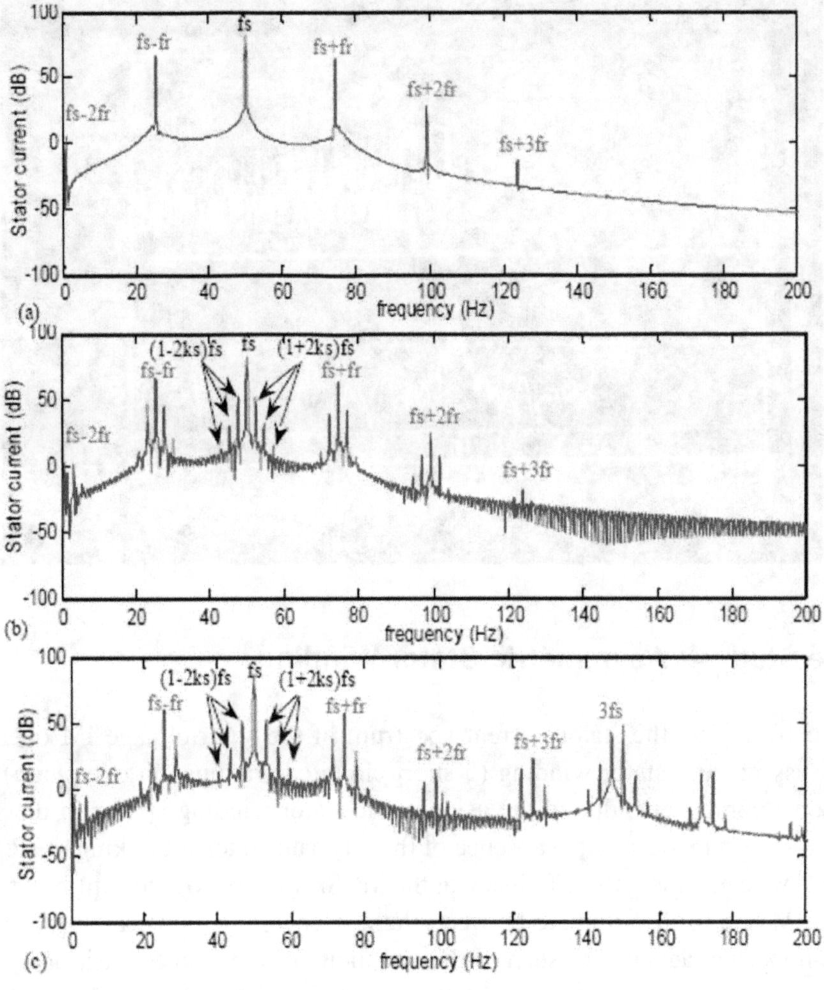

8 (c), frequencies due to load variation, rotor asymmetry and unbalanced supply voltage appeared simultaneously in the stator current spectrum and also Figure 9 shows Stator current spectrum for motor operating under variable load condition in Agilent DSOX-2014A with broken rotor bars.

Figure 9. Stator current spectrum for motor operating under variable load condition in Agilent DSOX-2014A with broken rotor bars

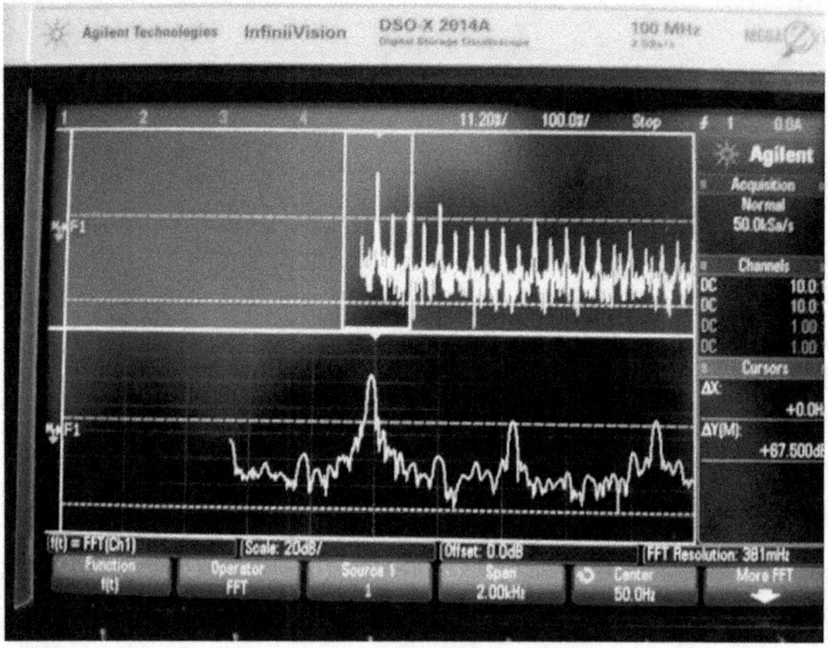

Case Study 4: Asymmetric Stator Winding

Figure 10 depicts the stator current spectrum of the squirrel cage IM operating under asymmetric stator winding (5 short-circuited windings in one phase). The harmonic components due to the asymmetric stator winding appear in the stator current spectrum, despite the absence of the mechanical load. Looking to all given figures, we can remark the efficiency of the MCSA method to detect different faults in the IM and to discriminate between them. Hence, this classical approach has some important advantages such as the simplicity of the data acquisition systems and the required software, along with the robustness of the tool, which has hitherto provided quite satisfactory results. Also, from the stator current spectrum analysis it is possible to detect rotor as well as stator faults. However, its validity has some drawbacks when the approach is applied under certain conditions.

Figure 10. Experimental stator current spectrum of unloaded IM, operating under asymmetric stator winding (5 short-circuited-windings in one phase)

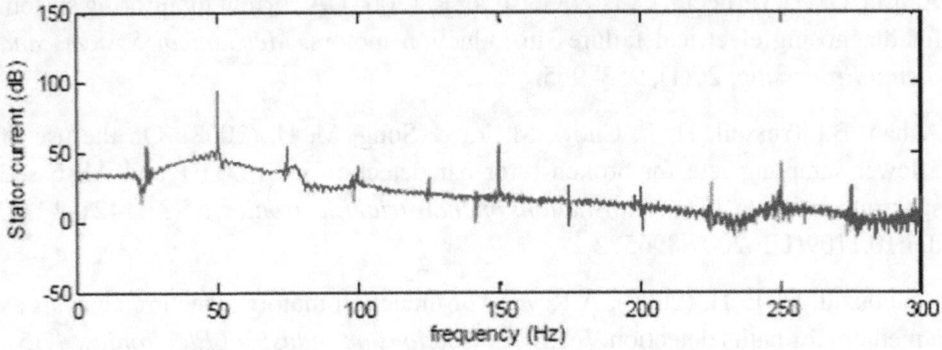

This is the case, for instance, of light-loaded or unloaded machines. In those situations, the slip *s* is very low and the sideband components practically overlap the supply frequency. This makes it difficult to detect their presence and to use them for the diagnosis, as Didier et al. (2007) remarked. Also, the frequency characterizing the fault must be computed considering the motor poles number, the slip, and the motor features in some cases. Also, another handicap of this approach is that it is not possible to relate the fault severity with the amplitude of these frequencies characterizing the fault because the amplitude depends on many factors such as the external load and the measurement noise shown in Bulent Ayhan (2008) and Alka Thakur (2013).

CONCLUSION

A review of the most frequent IM failures and their detections via the MCSA is presented. Initially, the phenomenon of asymmetrical rotor is presented and all the characteristic frequencies occurring in the stator current of IM due to cracked or broken bars are explained. Similarly, the phenomena of shorted turns in the stator windings, unbalanced stator voltage and load variation, in healthy motor and in the case of broken rotor bars, are studied. Experiment and simulation results showed the efficiency of the proposed method to spot and discriminate between the different faults but the major drawback of this method, is in the case of light-loaded or unloaded machines it is difficult to detect faults and fortunately an IM operates most of the time under its rated load torque.

REFERENCES

Acosta, G. G., Verucchi, C. J., & Gelso, E. R. (2004). A current monitoring system for diagnosing electrical failures in induction motors. *Mechanical Systems and Signal Processing*, *20*(1), 953–965.

Ayhan, B., Trussell, H. J., Chow, M. Y., & Song, M. H. (2008). On the use of a lower sampling rate for broken rotor bar detection with DTFT and AR-based spectrum methods. *IEEE Transactions on Industrial Electronics*, *55*(3), 1421–1434. doi:10.1109/TIE.2007.896522

Benbouzid, M. E. H. (2000). A Review of induction motors signature analysis as a medium for faults detection. *IEEE Transactions on Industrial Electronics*, *47*(5), 984–993. doi:10.1109/41.873206

Blodt, M., Chabert, M., Regnier, J., & Faucher, J. (2006). Mechanical Load Fault Detection in Induction Motors by Stator Current Time-Frequency Analysis. *IEEE Transactions on Industry Applications*, *42*(6), 1454–1463. doi:10.1109/TIA.2006.882631

Çalis, H., & Çakir, A. (2007). Rotor bar fault diagnosis in three phase induction motors by monitoring fluctuations of motor current zero crossing instants. *Electric Power Systems Research*, *77*(5), 385–392. doi:10.1016/j.epsr.2006.03.017

Casimir, R., Boutleux, E., Clerc, G., & Yahoui, A. (2006). The use of features selection and nearest neighbours rule for faults diagnostic in induction motors. *Engineering Applications of Artificial Intelligence*, *19*(2), 169–177. doi:10.1016/j.engappai.2005.07.004

Didier, G., Ternisien, E., Caspary, O., & Razik, H. (2007). A new approach to detect broken rotor bars in induction machines by current spectrum analysis. *Mechanical Systems and Signal Processing*, *21*(2), 1127–1142. doi:10.1016/j.ymssp.2006.03.002

Robinson, J., Whelan, C. D., & Kent Haggerty, N. (2004). Trends in advanced motor protection and monitoring. *IEEE Transactions on Industry Applications*, *40*(3), 853–860. doi:10.1109/TIA.2004.827472

Thakur, A., Wadhwani, S., & Sondhiya, V. (2013). Health monitoring of rotating electrical machine using soft computing techniques: A Review. *International Journal of Scientific and Research Publications*, *3*(11), 1–3.

Chapter 5
Application of Statistical Quality Control in Testing Labs for Estimation of Machine Interference

Nitin K. Mandavgade
Nagpur Institute of Technology, India

Santosh B. Jaju
G. H. Raisoni College of Engineering Nagpur, India

Ramesh R. Lakhe
Shreyas Quality Management System, India

ABSTRACT

The performance and maintenance of testing laboratories is a prime issue. The quality of coal test results not only depends on performance of individual results but it also depends on the performance of various tests in the same laboratory. Machine interference is a significant problem in many manufacturing system and testing equipment. The variation of results for testing equipment may be due to various factors which need to calculate the uncertainty of measurement to show the accuracy of the machine. In case of coal testing laboratory, the plant layout and surrounding environment affects the performance of the system. The machine interference comes under variable causes which may affect the result. This chapter proposes a methodology for constructing system performance measures, finding out the various factors responsible for variations in result. The chapter deals with estimation of machine interference existence using variable control chart approach for coal testing equipment. The analysis of results for such machine interference will be useful and significant for system designers and practitioners.

DOI: 10.4018/978-1-5225-5406-6.ch005

INTRODUCTION

In nature two extremely similar things are difficult to obtain. If at all we come across exactly similar results it must be only by chance. This fact holds good for testing process as well. No testing process is good enough to produce an exactly similar result for the same sample on same equipment with same operator. The testing process gets affected by various sources such as quantitative as well as qualitative. The consistency of result gets affected by sample, operator and environmental conditions etc. The result of test process gets affected after controlling all above factors which may be due to machine interference.

A statistical quality control procedure is an important tool in total system of quality control. The purpose of this discussion is to outline a simple statistical procedure that is widely applicable and practical in testing laboratory (Westgard, Barry, & Hunl, 1981).

In case of material testing, the result of a measurement is only an approximation to the value of the measurand and is only complete when it is accompanied by the uncertainty. Computing the reported test result is straightforward, however, computing the uncertainty associated with the test result requires more consideration. The uncertainty evaluation process will encompass a number of influencing quantities that affect the result obtained for the measurand. In order to quantify the uncertainty we will have to consider all the factors that could influence the results. Material testing measurement process is always doubtful about the value of quantity to be measured which provides the basis for safe working of components during their operations. Therefore, it is essential that measurement professionals explain to their clients the value of their work and its limitations.

A reading will give the false impression that the measurement is accurate. As there are other variables which would be accounted for, critical applications those will rely upon if measurements might have errors. Those experienced with measuring equipment are typically aware of this fact and often provide a guess of a reading to the best of their knowledge. But skill and experience often plays a role in how accurate those guesses might be. However, as the benefits of measurement uncertainty (UOM) are becoming more widely known, the practice is being put to use with greater credibility in many critical fields.

Because a realistic uncertainty measurement can account for actual variables that may be encountered, possible solutions can be provided and procedures can be enacted to account for those variables. A realistic uncertainty reading also lends greater credibility to the measurement information being produced. International standards and procedures have been created that outlines how to take a realistic measurement uncertainty reading utilizing a tool known as the "Uncertainty Budget" (Mandavgade et al., 2011b).

Variations due to chance causes are inevitable in any process or product. They are difficult to trace and difficult to control even under best conditions of testing. It has been established that if the variations are due to chance factors alone, the observations will follow a 'normal curve.' Knowledge of behavior chance variation is the foundation on which the control chart analysis rests (Mahajan, 2001).

IDENTIFICATION OF FACTORS

The uncertainty of measurement gets affected by qualitative factors as well as quantitative factors. The effects of quantitative factors can be determined using analytic approach and data from calibration certificates, instruments manual, etc. (Mandavgade et al 2011a). In case of qualitative terms, UOM is affected by various terminologies applicable to that particular case. According to the expert suggestions or opinions, questions filled by different people in particular field and practical experience of the author in field, some of the qualitative factors applicable for M/s. S.K. Mitra coal testing laboratory, Nagpur (India) are as below (Figure 1):

- **Operator:** The operator will affect the performance of test in terms of following factors.
    - **Field Knowledge of Operator:** If the operator is not skilled enough or if he lacks in the knowledge of a particular instruments and its procedure then the testing results may not be so accurate. The effects because of the field knowledge of operator are not possible to calculate in numerical terms.
    - **Experience of the Operator:** Experience plays a vital role in case of testing and measuring procedure. Experience somewhat related to field of knowledge, frequency of testing of the same sample. More the experience of the operator less will be chances of mistake. The effects because of the experience of operator are again impossible to calculate in analytical way.
    - **Confidence of the Operator:** The knowledge of the instruments increases the confidence of the operator for handling the test. Confidence plays a vital role in success to avoid the errors in procedure.
    - **Attitude of the Operator:** The positive attitude always reaches to success of everything. In field, if the operator is having the attitude to do the work by heart then the chances of errors will be rare or less. Attitude of the operator is a logical thing so hard to express in term of analytic value.

 - **Training of the Operator:** Before handling the instruments, training is must for the operator. Training enhances the working skill of the operator, increases the confidence and knowledge of the operator. All above parameters are related to the operator and varies from person to person.
- **Environmental Condition:** The performance of the test will be affected by some environmental condition as below.
  - **Location of Laboratory:** The location of laboratory affects the performance of final result depending upon the nature of product to be tested. If the location of laboratory is at first or second floor then the leveling of the foundation effects may occur. Generally all testing laboratories are preferred at ground floor with the proper foundation.
  - **Plant Layout:** Plant layout plays a role in increasing the efficiency of the operator. If the machines are arranged in zig-zag manner then it may have an effect in psychological performance of the operator.
  - **Temperature of the Surrounding:** Every test is to be performed at specific temperature as per manufacture's standard. But the temperature is very difficult to maintain in laboratory due to various reasons such as frequent entry of people, working efficiency of the cooling equipment etc.
  - **Humidity of the Surrounding:** Humidity will greatly affect the performance of coal as well as transformer oil. Humidity should be kept under control and measured properly.
  - **Wind Force:** Wind force will affect the performance of the test in hill station area where force of wind is so high. A wind force effect depends on the position of the lab and also to direct exposure to the environment.
  - **The Number of People in Lab:** The frequent entry and exit of number of people in laboratory increase internal heat of lab and healthy working environment which affect the performance of measuring and testing.
  - **Heating Equipment in Laboratory:** The operation timing of heating equipment affects the performance of nearby working equipment.
- **Sample Preparation:** The sample affects the test performance in the following terms
  - **Homogeneity of the Sample:** In case of coal and oil, homogeneity of the sample is most important factor affecting the consistency of the result. While taking the sample reading, sub sample should be taken from the same sample.
  - **Storage Condition of Sample:** Storage condition of sample is again more important in case of coal and oil. If transformer oil is not stored in air tight condition then changes in oil properties may occur.

- **Preparation of Sample:** Preparation of sample in terms of quantity and quality is more important. The cleaning of glasswares should be done with the same sample. Then proper quantity of sample is to be taken for testing.
- **Exposure to the Environment:** If the sample is kept open and exposed to the environment then the properties of sample like coal and oil may change.

- **Machine Interference:** The effect of others machines on the performance of individual equipment when all equipment are in working condition at the same time. A number of tests are performed in a single laboratory. The performance of individual affects the globe, which in turn may affect the performance of others in very small quantity (Figure 1).

METHODOLOGY

The methodology used to find out the effects of individual equipment on every test performed at S. K. Mitra coal testing laboratory, Nagpur is as shown in Figure 2. The numbers of tests are performed on the single equipment. The single sample of

Figure 1. Fish bone diagram for sources of variation in result

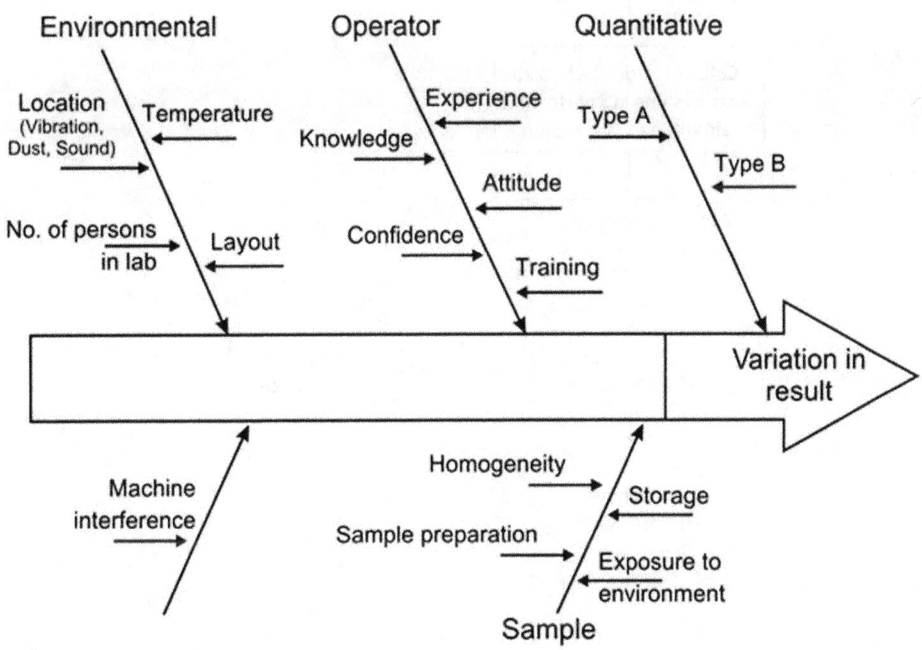

Figure 2. Methodology for machine interference

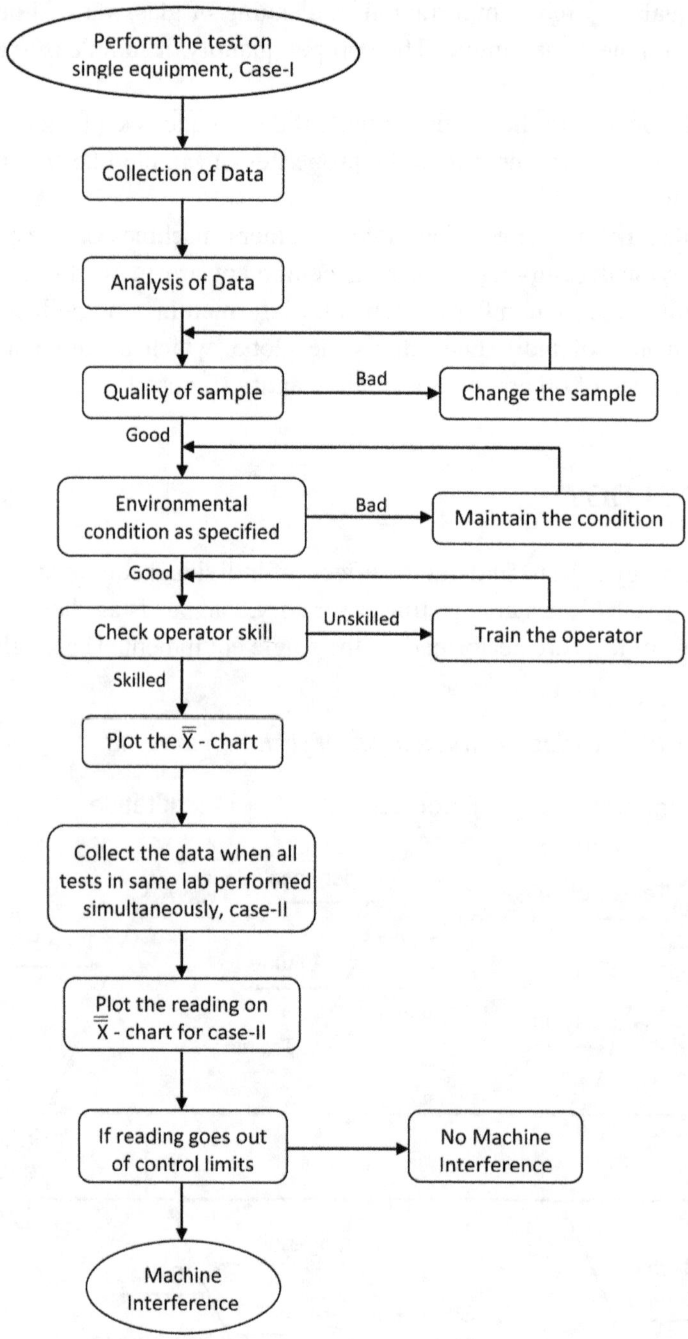

coal gets divided into five subgroups and again making four sub sample from each subgroup. The complete data is subdivided into six zones. The expected frequency for each zone is calculated using random numbers. The expected frequency and observed frequency are compared with each other and results show that they follow normal distribution shape.

In case of S. K. Mitra coal laboratory, various tests of coal such as moisture, ash, volatile, sulphur and gross calorific value of coal are performed. Every individual test is performed separately and the data is collected for at least five subgroup sample each containing four samples. The variable control charts technique is used to check the effect of machine interference (Benmeyan, 1998). The data of at least twenty samples are used to plot upper and lower control limits of X-chart. The test on the same equipment with same sample are again performed when all other equipment in the laboratory is in working condition taking care that environmental conditions

Figure 3. Lab layout of S.K. Mitra coal testing lab

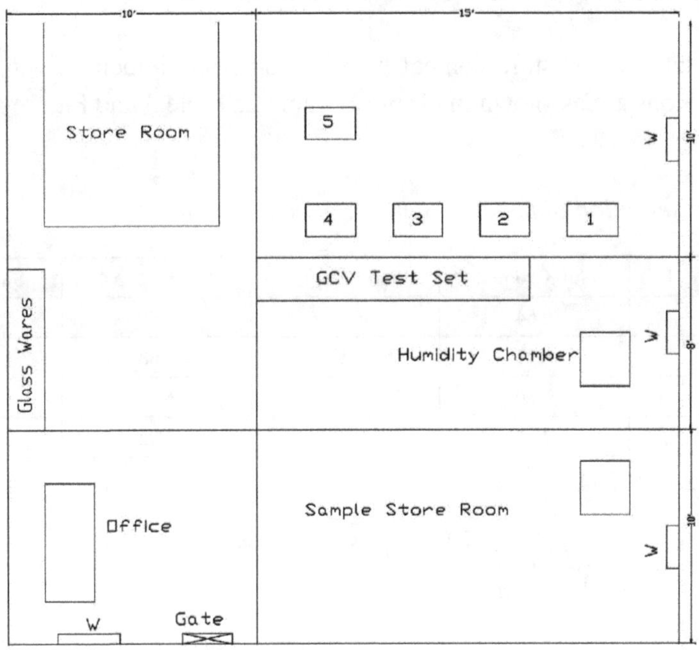

Notifications used-
1, 2,3,4,5 – Furnaces of different ranges
GCV - Gross calorific value test set
W- Window of the room.
Gate- - Entrance of the laboratory.

are as specified. The data for the next ten readings are plotted on the same control chart. If any reading goes out of the control limits, it means that the measurement is affected by impact of machine interference because all other factors are in control conditions (Figure 2).

OBSERVATION

The properties of coal are tested in terms of air dried moisture, ash content, volatile matter, gross calorific value and sulphur content of coal. The data collected for the samples used in thermal power station of India tested at M/s. S.K. Mitra, coal testing laboratory, Nagpur is presented.

The detail laboratory layouts are shown in Figure 3. The layout shows the detailed position of various equipment and storage condition. While performing the individual test we have controlled all above parameters as describe in Figure 1 (Mahajan 2001).

Measurement for Air Dried Moisture

1. The data for moisture content when all other equipment is not in working conditions are as shown in (Table 1, Table 2, Table 3 and Figure 4).

Table 1. Observation of moisture content of coal in ppm

Sub Group 1	Sub Group 2	Sub Group 3	Sub Group 4	Sub Group 5
6.82	6.64	6.93	6.80	6.69
6.53	6.97	6.58	6.58	6.88
6.26	6.60	6.50	6.81	6.84
6.63	7.15	6.78	7.02	6.90

Average = $\overline{x} = 6.7545$

$$Std.deviation = \sqrt{\frac{\Sigma\left(x_i - \overline{x}\right)^2}{n}} = \sqrt{\frac{0.83+1}{20}}$$

σ=0.204

$$Zone = \frac{Max^m - Min^m}{Number\ of\ zone\ to\ be\ made}$$

$$= \frac{7.15 - 6.26}{6}$$

=0.148

Then 0.148 to be added to each value

Table 2. Zone frequency

X	$Z = \frac{x - \overline{x}}{\sigma}$	Random No
6.26	-2.11	0.0174
6.408	-1.49	0.0681
6.556	-0.864	0.1949
6.704	-0.237	0.4090
6.852	0.389	0.6480
7.00	1.017	0.8438
7.15	1.653	0.9505

Total area under curve
T_1= Maximum Random Number Value – Minimum Random Number Value
= 0.9505-0.0174
= 0.9331

Table 3. Frequency zone table

Zones	Observed Frequency (O_f)	Expected Frequency (e_f)
6.25-6.408 (Z_1)	1	1.086
6.408-6.556 (Z_2)	2	2.71
6.556-6.704 (Z_3)	6	4.58
6.852-6.852 (Z_4)	5	5.122
6.852-7.00 (Z_5)	4	4.196
7.00-7.15 (Z_6)	2	2.287

$$e_{f(1)} = \frac{R_2 - R_1}{T_1} * n = \frac{0.0681 - 0.0174}{0.9331} * 20 = 1.086$$

$$e_{f(2)} = \frac{R_3 - R_2}{T_1} * n = \frac{0.1949 - 0.0681}{0.9331} * 20 = 2.71$$

$$e_{f(3)} = \frac{R_4 - R_3}{T_1} * n = \frac{0.4090 - 0.1949}{0.9331} * 20 = 4.58$$

$$e_{f(4)} = \frac{R_5 - R_4}{T_1} * n = \frac{0.6480 - 0.1949}{0.9331} * 20 = 5.122$$

$$e_{f(5)} = \frac{R_6 - R_5}{T_1} * n = \frac{0.8438 - 0.6480}{0.9331} * 20 = 4.196$$

$$e_{f(6)} = \frac{R_7 - R_6}{T_1} * n = \frac{0.9505 - 0.8438}{0.9331} * 20 = 2.287$$

Figure 4. Normal distribution curve for moisture data of coal

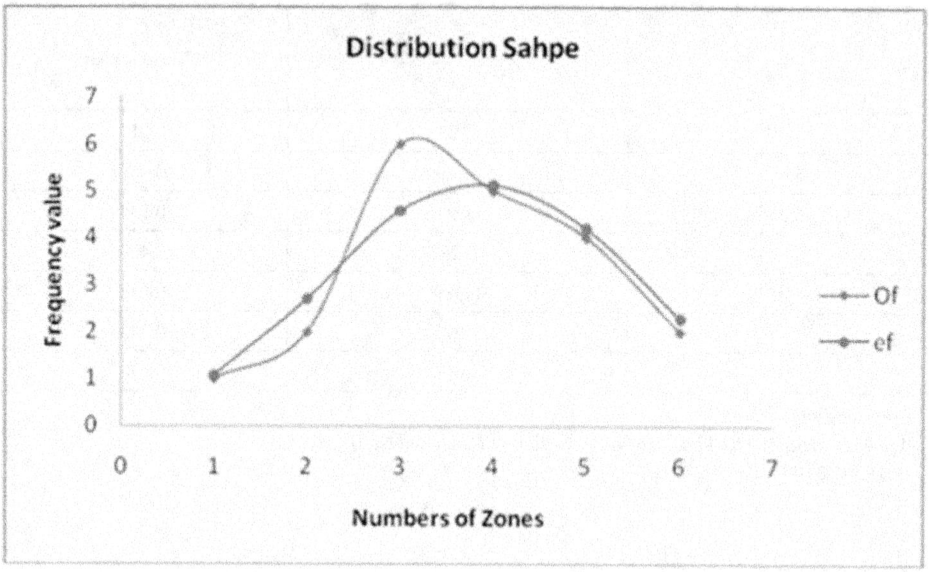

2. Same test performed again when all other equipment in laboratory are in working conditions. The observation data are as shown in Table 4. The $\bar{x}$ -chart for the tan delta data is as shown in Figure 5.

Table 4. Data of moisture of coal when all equipment in lab are in working conditions

Sr. No	1	2	3	4	5	6	7	8	9	10
Value	6.42	6.92	6.72	6.54	6.83	6.89	6.94	6.68	6.32	6.53

Average = $\bar{x}$ = *30.718*

The averages of each sub sample are as calculated below.

X_1= 6.56, X_2= 6.84, X_3=6.69, X_4=6.80, X_5=6.82

$\bar{\bar{x}} = 6.74$

Ranges-

R_1= 0.56, R_2=0.55, R_3= 0.43, R_4= 0.44, R_5=0.21

$\bar{R} = 0.43$

For $\bar{x}$ -Chart:

$$\text{UCL}\,\bar{x} = \bar{\bar{x}} + A_2\bar{R}$$

= 6.74 + 0.73(0.43)

= 7.053

$$\text{LCL}\,\bar{x} = \bar{\bar{x}} - A_2\bar{R}$$

6.74 -0.73(0.43)

= 6.426.

Figure 5. $\overline{x}$ *-Chart for moisture content data of coal*

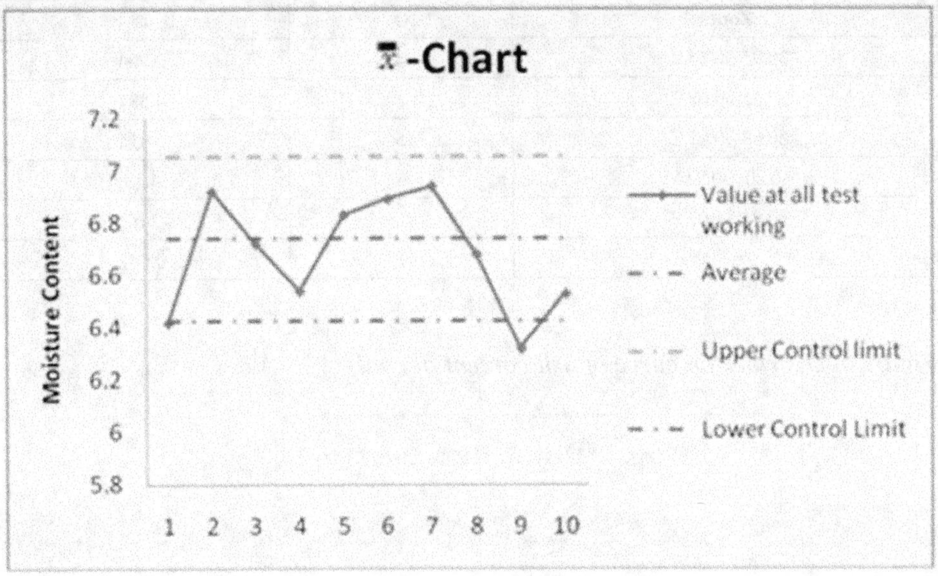

From the graph (Figure 5) as point number nine goes below the lower control limit it means there is an existence of machine interference for variation in result. The environmental conditions, quality, sample and operator are controlled in this case. Therefore, the variation is only because of machine interference.

The Measurement of Ash Content

1. The data for ash content when all other equipment in laboratory are not in working conditions (Table 5, Table 6, Figure 6).
2. Same test performed again when all other equipment in laboratory are in working conditions. The data is collected as in (Table 7, Figure 7).

Table 5. Observation of ash content of coal

Sub Group 1	Sub Group 2	Sub Group 3	Sub Group 4	Sub Group 5
30.75	30.97	30.69	30.62	30.75
30.91	31.21	30.42	30.47	30.62
30.98	30.80	30.45	30.52	30.58
30.91	30.86	30.58	30.60	30.72

Table 6. Frequency distribution zone

Zones	O_f	e_f
30.92-30.55 Z_1	4	2.904
30.55-30.68 Z_2	5	4.887
30.68-30.81 Z_3	5	5.314
30.80- 30.94 Z_4	3	4.13
30.94-31.07 Z_5	2	2.07
31.07-31.2 Z_6	1	0.687

Figure 6. Distribution curve of ash content of coal

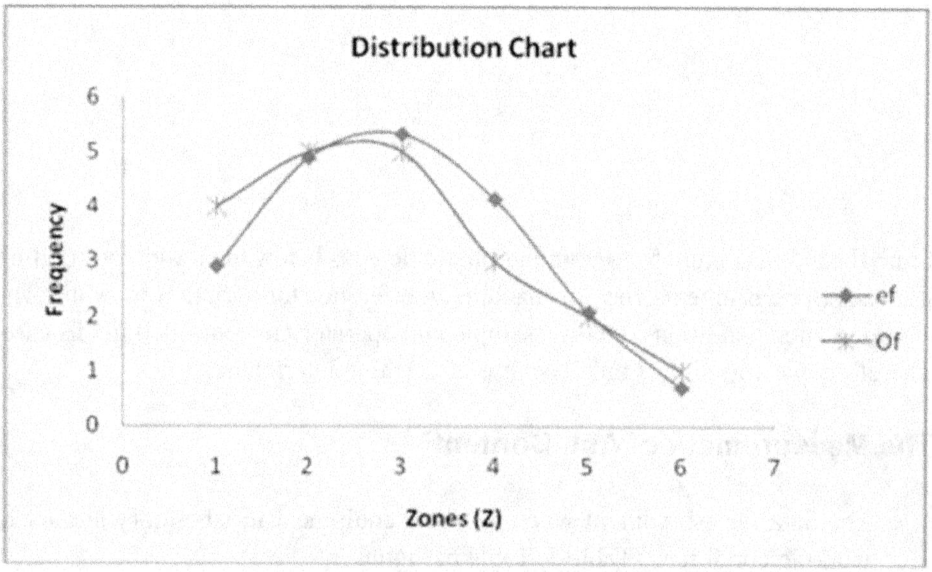

Table 7. Data ash content of coal when all equipments in lab are working conditions

Sr. No.	1	2	3	4	5	6	7	8	9	10
Value	30.89	30.58	30.65	30.85	30.58	30.55	30.82	30.77	30.49	30.48

Figure 7. $\bar{x}$ -Chart for ash content of coal

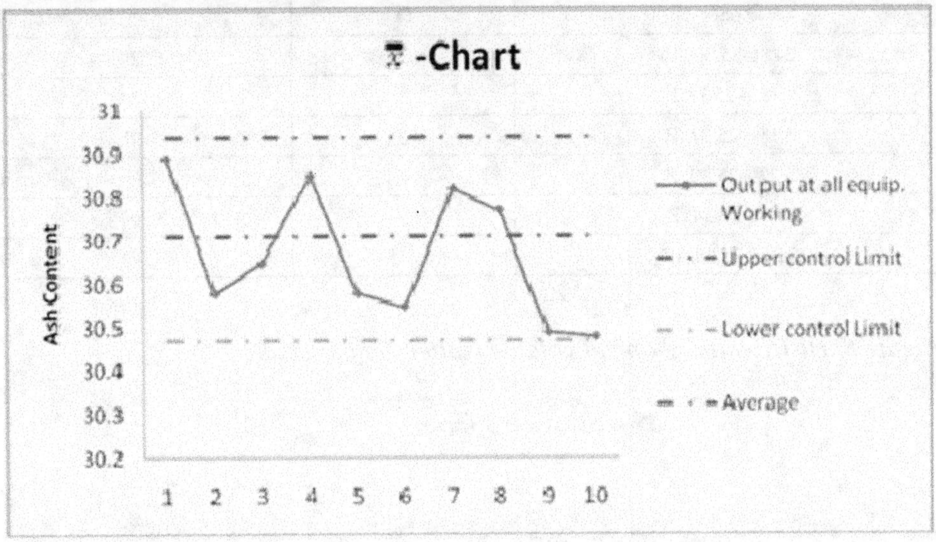

The Measurements of Volatile Matter

1. The data for volatile content of coal when all other equipment is not in working conditions (Table 8, Table 9 and Figure 8).
2. The same test performed again when all other equipment in laboratory are in working conditions. The data is collected as follow (Table 10, Figure 9).

Table 8. Observation of volatile matter in coal

Sub Group 1	Sub Group 2	Sub Group 3	Sub Group 4	Sub Group 5
24.86	25.11	25.18	25.25	25.30
25.42	25.42	25.42	25.35	24.99
25.58	25.37	25.26	25.20	25.12
25.38	25.68	25.22	25.01	25.19

Average= $\overline{X}$ *=25.27*

Table 9. Frequency zone table

Zones	O_f	e_f
24.86-25.00(Z_1)	2	1.327
25.00 – 25.13(Z_2)	3	3.184
25.13 – 25.27 (Z_3)	6	5.48
25.27 – 25.40 (Z_4)	4	5.15
25.40 - 25.54 (Z_5)	3	3.508
25.54 – 25.68s (Z_6)	2	1.35

Figure 8. Distribution shape of volatile matter data of coal

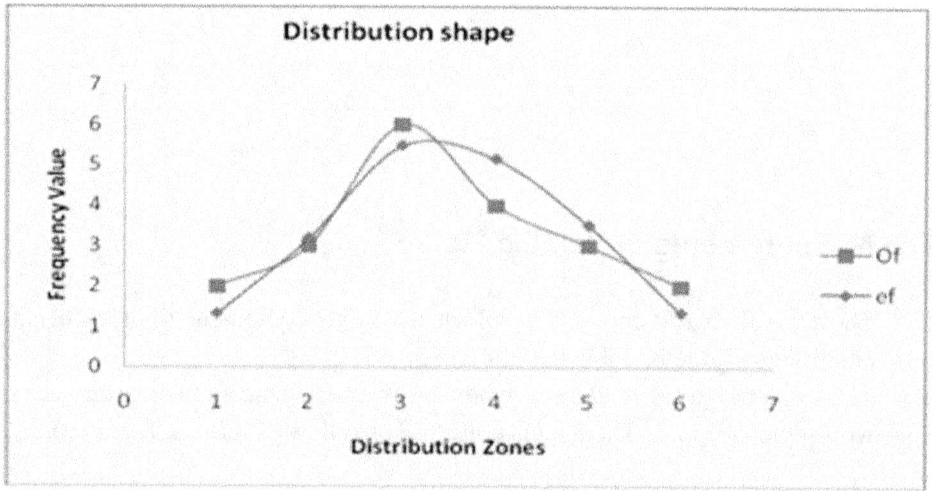

Table 10. Data of volatile content of coal when all equipments in lab are working conditions

Sr. No.	1	2	3	4	5	6	7	8	9	10
Value	25.29	25.03	24.90	24.98	25.12	25.22	25.13	24.86	24.85	24.95

Figure 9. $\overline{x}$ *-Chart of volatile data content of coal*

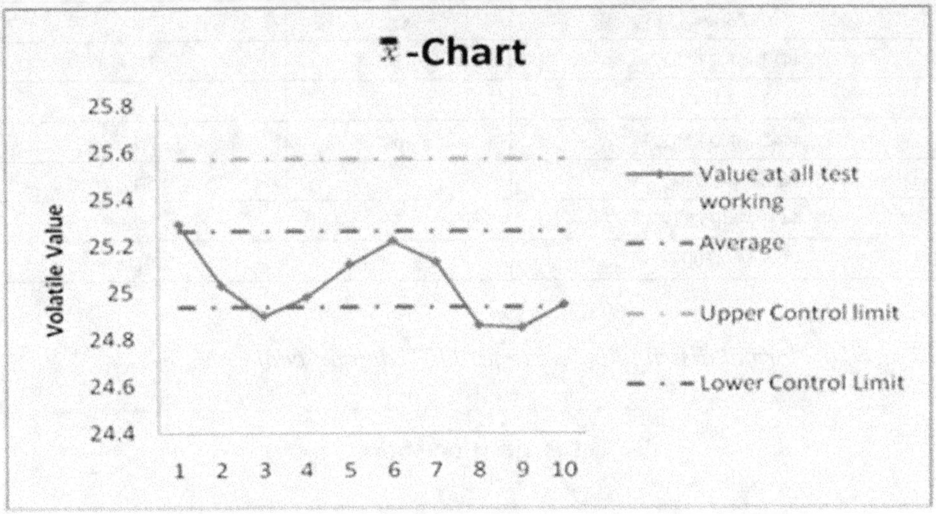

The Values of Gross Calorific Value of Coal

1. The data of gross calorific value of coal when all other equipment are not in working conditions (Mandavgade, Jaju, & Lakhe, 2011b) (Table 11, Table 12, Figure 10).
2. The same test performed again when all other equipment in laboratory are in working conditions. The data are collected as follows (Table 13, Figure 11).

The Values of Sulphur Content of Coal

1. The data of Sulphur content of coal when all other equipment are not in working conditions (Table 14, Table 15, Figure 12).

Table 11. Observation of GCV of coal

Sub Group 1	Sub Group 2	Sub Group 3	Sub Group 4	Sub Group 5
4413	4472	4462	4435	4469
4500	4448	4475	4431	4425
4456	4465	4454	4438	4492
4445	4449	4484	4452	4449

Table 12. Frequency zone table

Zones	O_f	e_f
4413 – 4427.5 (Z_1)	2	1.49
4427.5 – 4442 (Z_2)	3	3.48
4442 – 4456.5 (Z_3)	6	5.17
4456.5 – 4471 (Z_4)	4	5.17
4471 - 4485.5 (Z_5)	3	3.28
4485.5 – 4500 (Z_6)	2	1.4

Figure 10. Normal distribution curve for GCV data of coal

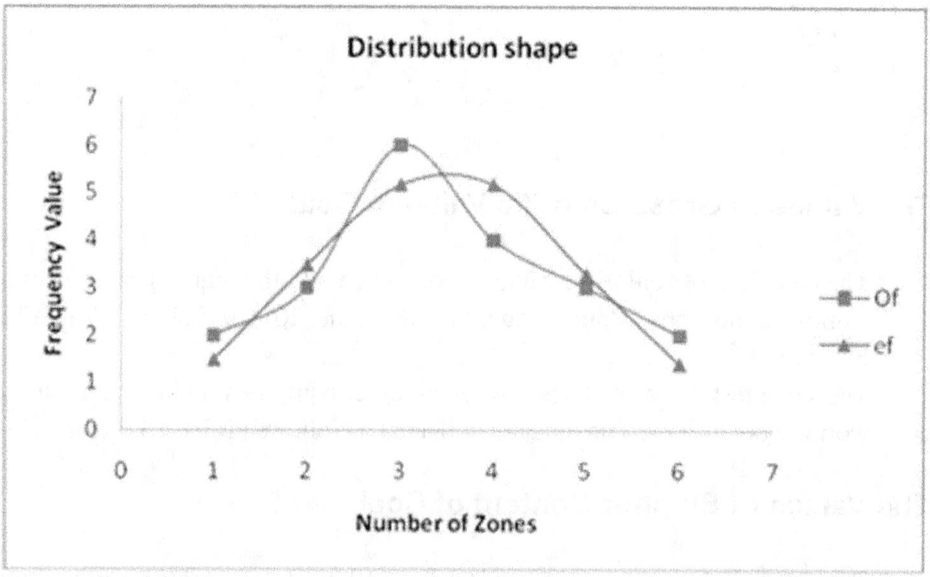

Table 13. Data of GCV of coal when all equipment in lab are working conditions

Sr. No.	1	2	3	4	5	6	7	8	9	10
Value	4448	4485	4498	4468	4478	4452	4440	4439	4468	4485

Figure 11. $\overline{x}$ -chart for GCV of coal

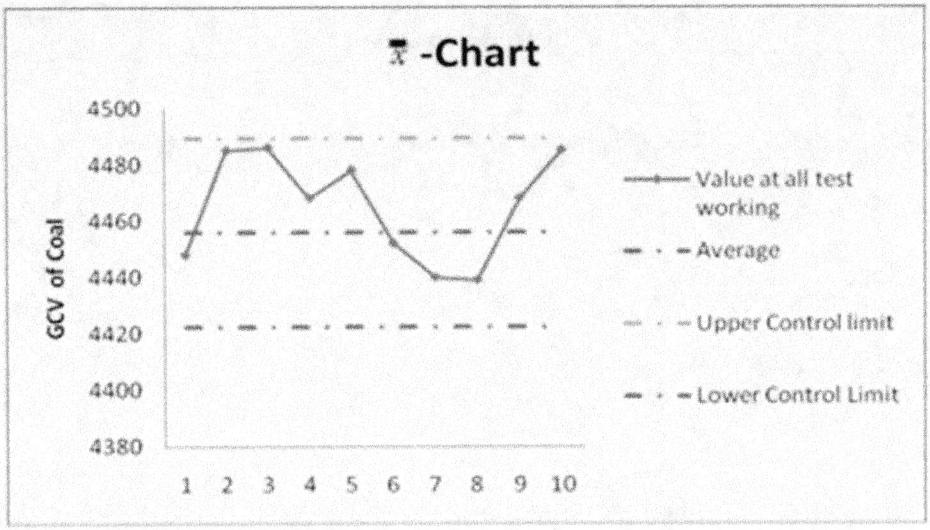

Table 14. Observations of sulphur content data of coal in %

Sub Group 1	Sub Group 2	Sub Group 3	Sub Group 4	Sub Group 5
0.550	0.460	0.550	0.502	0.520
0.500	0.480	0.490	0.510	0.495
0.480	0.480	0.485	0.512	0.515
0.460	0.530	0.495	0.520	0.512

Average= $\overline{X}$ = 0.5023

Table 15. Frequency zone table

Zones	O_r	e_r
0.46 – 0.475 (Z_1)	1.92	2
0.475 – 0.49 (Z_2)	3.75	4
0.49 – 0.505 (Z_3)	5.10	5
0.505 – 0.52 (Z_4)	4.77	4
0.52 – 0.535 (Z_5)	3.11	3
0.535 – 0.55 (Z_6)	1.34	2

Figure 12. Distribution curve for sulphur content data of coal

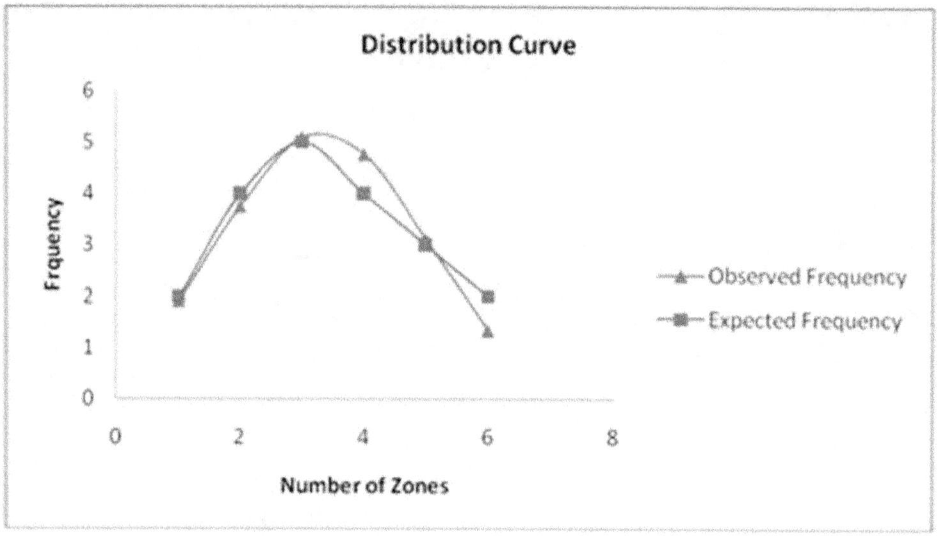

2. The same test performed again when all other equipment in laboratory are in working conditions. The data are collected as follows (Table 16, Figure 13).

DISCUSSION AND CONCLUSION

The quality of coal used in thermal power station of Koradi and Khaperkheda, Nagpur is checked in term of moisture content, ash content, volatile matter, sulphur content and gross calorific value of coal. The uncertainty in measurement of these terms gets affected by various sources. The paper discussed the various possible sources responsible for variation in the results of test. The statistical quality technique is used to find out the existence of machine interference in case S.K. Mitra coal testing laboratory, Nagpur (India). From the control chart for moisture, first and ninth readings are outside the control limit. In case of control chart for volatile matter third, eighth

Table 16. Data of sulphur content of coal when all equipment in lab are working conditions

Sr. No.	1	2	3	4	5	6	7	8	9	10
Value	0.530	0.524	0.498	0.510	0.505	0.498	0.516	0.520	0.535	0.530

Figure 13. $\overline{x}$ *-chart for sulphur content of coal*

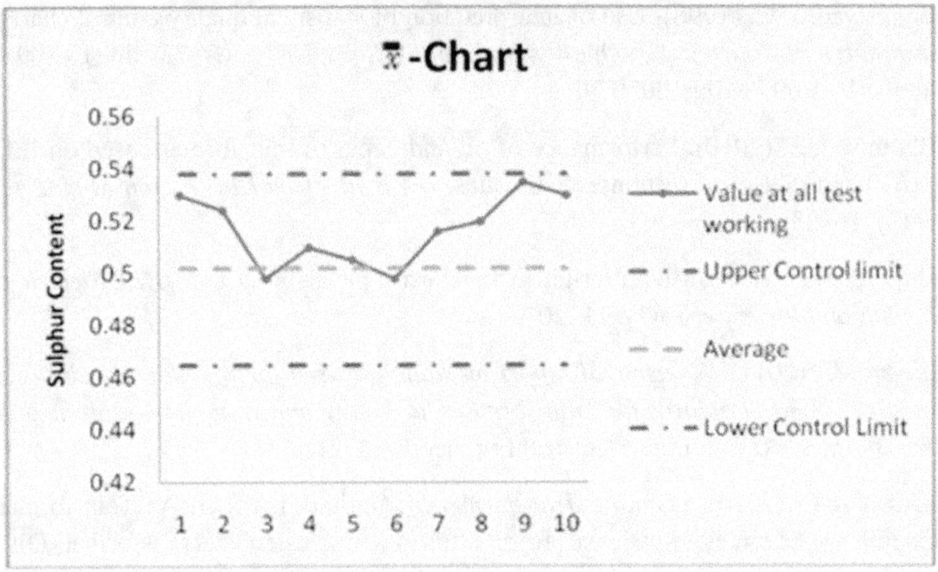

and ninth readings fall outside the control limits. The deviation for control limits is found when all equipment in laboratory are in working condition simultaneously. Therefore, from the result, only two tests such as moisture content and volatile matter test are affected by machine interference, since all other conditions are in control.

Precaution

The precautions to be taken while performing the tests in central oil testing lab are as follows.

- While performing the moisture test and volatile matter test, all other equipment in the same laboratory should not be in working conditions.
- The environmental conditions should be maintained as specified.

ACKNOWLEDGMENT

The author acknowledged authority of M/S. S.K. Mitra coal testing laboratory, Nagpur (India) for giving the opportunity to carry out the research work.

REFERENCES

Benmeyan, J. C. (1998). Use of interpretation of statistical quality control charts. *International Journal for Quality in Health Care*, *10*(1), 69–73. doi:10.1093/intqhc/10.1.69 PMID:10030790

Bhumiwat, S. (2010). Performance of oil and paper in transformer based on IEC 61620 and dielectric response techniques. *IEEE Electrical Insulation Magazine*, *26*(3), 16–23.

Du, Y. (1999). Moisture equilibrium in transformer paper-oil system. *IEEE Electrical Insulation Magazine*, *15*(1), 11–20.

Edvard, C. (2011). *A technical report on standard used for transformer oil tests. In 2011 ASTM Standards for Transformer Oil Testing and Complete Analysis of a Transformer's Oil.* Online Electrical Engineering Portal.

I.A.R.GRAY. (2010). *A Guide to Transformer Oil Analysis*. I.A.R.GRAY Transformer Chemistry Services. Retrieved from http://www.satcs.co.za/Transformer_Oil_Analysis.pdf

Mahajan, M. S. (2001). *Statistical quality control* (3rd ed.). New York, NY: McGraw-Hill.

Mandavgade, N. K., Jaju, S. B., & Lakhe, R. R. (2011a). Determination of uncertainty in gross calorific value of coal using Bomb calorimeter. *International Journal of Measurement Technologies and Instrumentation*, *1*(4), 45–52. doi:10.4018/ijmtie.2011100104

Mandavgade, N. K., Jaju, S. B., & Lakhe, R. R. (2011b). Mathematical modeling of effects of various factors on uncertainty of mea-surement in material testing. *Proceedings of the International Conference on Mechanical Engineer-ing and Technology*.

Mandavgade, N. K., Jaju, S. B., & Lakhe, R. R. (2012). Evaluation of uncertainty of measurement of automatic tan delta and resistivity test set for transformer oil. *Int. J. Metrol. Qual. Eng.*, *3*(1), 39–45. doi:10.1051/ijmqe/2012004

Mandavgade, N.K., Jaju, S.B., & Lakhe, R.R. (2012). Identifying uncertainty sources in measurements – a methodology. *Journal of Measurement Science and Instrumentation*, *3*(4), 307–312.

Stannett, A. W. (1951). The conductivity of hydrocarbon transformer oil containing water and solid conducting particles. *British Journal of Applied Physics*, *2*(4), 304. doi:10.1088/0508-3443/2/4/304

Stevens, P. (2008). The Technical Report on 'A Methodology for Assessing the Performance of National Oil Companies'. Academic Press.

Westgard, J. O., Barry, P. L., & Hunl, M. R. (1981). A multi rule shewart chart for quality control in clini-cal laboratory. *Clinical Chemistry*, *27*(3), 493–501. PMID:7471403

Westgard, J. O., Barry, P. L., & Hunl, M. R. (1981). A multi rule shewart chart for quality control in clinical laboratory. *Clinical Chemistry*, *27*(3), 493–501. PMID:7471403

APPENDIX

Table 17. Table of the standard normal cumulative distribution function $\Phi(z)$

z	0.00	0.01	0.02	0.03	0.04	0.05	0.06	0.07	0.08	0.09
-3.4	0.0003	0.0003	0.0003	0.0003	0.0003	0.0003	0.0003	0.0003	0.0003	0.0002
-3.3	0.0005	0.0005	0.0005	0.0004	0.0004	0.0004	0.0004	0.0004	0.0004	0.0003
-3.2	0.0007	0.0007	0.0006	0.0006	0.0006	0.0006	0.0006	0.0005	0.0005	0.0005
-3.1	0.0010	0.0009	0.0009	0.0009	0.0008	0.0008	0.0008	0.0008	0.0007	0.0007
-3.0	0.0013	0.0013	0.0013	0.0012	0.0012	0.0011	0.0011	0.0011	0.0010	0.0010
-2.9	0.0019	0.0018	0.0018	0.0017	0.0016	0.0016	0.0015	0.0015	0.0014	0.0014
-2.8	0.0026	0.0025	0.0024	0.0023	0.0023	0.0022	0.0021	0.0021	0.0020	0.0019
-2.7	0.0035	0.0034	0.0033	0.0032	0.0031	0.0030	0.0029	0.0028	0.0027	0.0026
-2.6	0.0047	0.0045	0.0044	0.0043	0.0041	0.0040	0.0039	0.0038	0.0037	0.0036
-2.5	0.0062	0.0060	0.0059	0.0057	0.0055	0.0054	0.0052	0.0051	0.0049	0.0048
-2.4	0.0082	0.0080	0.0078	0.0075	0.0073	0.0071	0.0069	0.0068	0.0066	0.0064
-2.3	0.0107	0.0104	0.0102	0.0099	0.0096	0.0094	0.0091	0.0089	0.0087	0.0084
-2.2	0.0139	0.0136	0.0132	0.0129	0.0125	0.0122	0.0119	0.0116	0.0113	0.0110
-2.1	0.0179	0.0174	0.0170	0.0166	0.0162	0.0158	0.0154	0.0150	0.0146	0.0143
-2.0	0.0228	0.0222	0.0217	0.0212	0.0207	0.0202	0.0197	0.0192	0.0188	0.0183
-1.9	0.0287	0.0281	0.0274	0.0268	0.0262	0.0256	0.0250	0.0244	0.0239	0.0233
-1.8	0.0359	0.0351	0.0344	0.0336	0.0329	0.0322	0.0314	0.0307	0.0301	0.0294
-1.7	0.0446	0.0436	0.0427	0.0418	0.0409	0.0401	0.0392	0.0384	0.0375	0.0367
-1.6	0.0548	0.0537	0.0526	0.0516	0.0505	0.0495	0.0485	0.0475	0.0465	0.0455
-1.5	0.0668	0.0655	0.0643	0.0630	0.0618	0.0606	0.0594	0.0582	0.0571	0.0559
-1.4	0.0808	0.0793	0.0778	0.0764	0.0749	0.0735	0.0721	0.0708	0.0694	0.0681
-1.3	0.0968	0.0951	0.0934	0.0918	0.0901	0.0885	0.0869	0.0853	0.0838	0.0823
-1.2	0.1151	0.1131	0.1112	0.1093	0.1075	0.1056	0.1038	0.1020	0.1003	0.0985
-1.1	0.1357	0.1335	0.1314	0.1292	0.1271	0.1251	0.1230	0.1210	0.1190	0.1170
-1.0	0.1587	0.1562	0.1539	0.1515	0.1492	0.1469	0.1446	0.1423	0.1401	0.1379
-0.9	0.1841	0.1814	0.1788	0.1762	0.1736	0.1711	0.1685	0.1660	0.1635	0.1611
-0.8	0.2119	0.2090	0.2061	0.2033	0.2005	0.1977	0.1949	0.1922	0.1894	0.1867
-0.7	0.2420	0.2389	0.2358	0.2327	0.2296	0.2266	0.2236	0.2206	0.2177	0.2148
-0.6	0.2743	0.2709	0.2676	0.2643	0.2611	0.2578	0.2546	0.2514	0.2483	0.2451
-0.5	0.3085	0.3050	0.3015	0.2981	0.2946	0.2912	0.2877	0.2843	0.2810	0.2776
-0.4	0.3446	0.3409	0.3372	0.3336	0.3300	0.3264	0.3228	0.3192	0.3156	0.3121
-0.3	0.3821	0.3783	0.3745	0.3707	0.3669	0.3632	0.3594	0.3557	0.3520	0.3483
-0.2	0.4207	0.4168	0.4129	0.4090	0.4052	0.4013	0.3974	0.3936	0.3897	0.3859
-0.1	0.4602	0.4562	0.4522	0.4483	0.4443	0.4404	0.4364	0.4325	0.4286	0.4247
-0.0	0.5000	0.4960	0.4920	0.4880	0.4840	0.4801	0.4761	0.4721	0.4681	0.4641
0.0	0.5000	0.5040	0.5080	0.5120	0.5160	0.5199	0.5239	0.5279	0.5319	0.5359
0.1	0.5398	0.5438	0.5478	0.5517	0.5557	0.5596	0.5636	0.5675	0.5714	0.5753
0.2	0.5793	0.5832	0.5871	0.5910	0.5948	0.5987	0.6026	0.6064	0.6103	0.6141
0.3	0.6179	0.6217	0.6255	0.6293	0.6331	0.6368	0.6406	0.6443	0.6480	0.6517
0.4	0.6554	0.6591	0.6628	0.6664	0.6700	0.6736	0.6772	0.6808	0.6844	0.6879
0.5	0.6915	0.6950	0.6985	0.7019	0.7054	0.7088	0.7123	0.7157	0.7190	0.7224
0.6	0.7257	0.7291	0.7324	0.7357	0.7389	0.7422	0.7454	0.7486	0.7517	0.7549
0.7	0.7580	0.7611	0.7642	0.7673	0.7704	0.7734	0.7764	0.7794	0.7823	0.7852
0.8	0.7881	0.7910	0.7939	0.7967	0.7995	0.8023	0.8051	0.8078	0.8106	0.8133
0.9	0.8159	0.8186	0.8212	0.8238	0.8264	0.8289	0.8315	0.8340	0.8365	0.8389
1.0	0.8413	0.8438	0.8461	0.8485	0.8508	0.8531	0.8554	0.8577	0.8599	0.8621
1.1	0.8643	0.8665	0.8686	0.8708	0.8729	0.8749	0.8770	0.8790	0.8810	0.8830
1.2	0.8849	0.8869	0.8888	0.8907	0.8925	0.8944	0.8962	0.8980	0.8997	0.9015
1.3	0.9032	0.9049	0.9066	0.9082	0.9099	0.9115	0.9131	0.9147	0.9162	0.9177
1.4	0.9192	0.9207	0.9222	0.9236	0.9251	0.9265	0.9279	0.9292	0.9306	0.9319
1.5	0.9332	0.9345	0.9357	0.9370	0.9382	0.9394	0.9406	0.9418	0.9429	0.9441
1.6	0.9452	0.9463	0.9474	0.9484	0.9495	0.9505	0.9515	0.9525	0.9535	0.9545
1.7	0.9554	0.9564	0.9573	0.9582	0.9591	0.9599	0.9608	0.9616	0.9625	0.9633
1.8	0.9641	0.9649	0.9656	0.9664	0.9671	0.9678	0.9686	0.9693	0.9699	0.9706
1.9	0.9713	0.9719	0.9726	0.9732	0.9738	0.9744	0.9750	0.9756	0.9761	0.9767
2.0	0.9772	0.9778	0.9783	0.9788	0.9793	0.9798	0.9803	0.9808	0.9812	0.9817
2.1	0.9821	0.9826	0.9830	0.9834	0.9838	0.9842	0.9846	0.9850	0.9854	0.9857
2.2	0.9861	0.9864	0.9868	0.9871	0.9875	0.9878	0.9881	0.9884	0.9887	0.9890
2.3	0.9893	0.9896	0.9898	0.9901	0.9904	0.9906	0.9909	0.9911	0.9913	0.9916
2.4	0.9918	0.9920	0.9922	0.9925	0.9927	0.9929	0.9931	0.9932	0.9934	0.9936
2.5	0.9938	0.9940	0.9941	0.9943	0.9945	0.9946	0.9948	0.9949	0.9951	0.9952
2.6	0.9953	0.9955	0.9956	0.9957	0.9959	0.9960	0.9961	0.9962	0.9963	0.9964
2.7	0.9965	0.9966	0.9967	0.9968	0.9969	0.9970	0.9971	0.9972	0.9973	0.9974
2.8	0.9974	0.9975	0.9976	0.9977	0.9977	0.9978	0.9979	0.9979	0.9980	0.9981
2.9	0.9981	0.9982	0.9982	0.9983	0.9984	0.9984	0.9985	0.9985	0.9986	0.9986
3.0	0.9987	0.9987	0.9987	0.9988	0.9988	0.9989	0.9989	0.9989	0.9990	0.9990
3.1	0.9990	0.9991	0.9991	0.9991	0.9992	0.9992	0.9992	0.9992	0.9993	0.9993
3.2	0.9993	0.9993	0.9994	0.9994	0.9994	0.9994	0.9994	0.9995	0.9995	0.9995
3.3	0.9995	0.9995	0.9995	0.9996	0.9996	0.9996	0.9996	0.9996	0.9996	0.9997
3.4	0.9997	0.9997	0.9997	0.9997	0.9997	0.9997	0.9997	0.9997	0.9997	0.9998

Chapter 6
Measurement Analysis of Ultra-Wide-Band and Ultra-Wide-Band-MIMO Antennas:
Review and Analysis

Madan Kumar Sharma
Rajasthan Technical University Kota, India

Mithilesh Kumar Kumar
Rajasthan Technical University Kota, India

Satya P. Singh
Nanyang Technological University, Singapore

ABSTRACT

This chapter mainly focused on the recent trends in the antenna design techniques for next-generation wireless communication systems. Ultra-wideband antenna and multi-input-multi-output antennas are very useful to achieve higher data rates. An antenna is a transducer that changes guided electromagnetic energy in a transmission line to radiated electromagnetic energy in free space. Antennas may also be observed as an impedance transformer, coupling among an input or line impedance, and the impedance of free space. The looming widespread commercial deployment of ultra-wideband (UWB) systems has flashed new interest in the subject of ultra-wideband antennas. The power levels approved by the FCC mean that every dB counts in a UWB system as much as or possibly even more so than in a standard narrowband system. Thus, in effect, UWB antenna is a precarious part of an overall UWB system design. Basic principles for ultra-wide-band (UWB) antenna design and radiation are presented and discussed in this chapter.

DOI: 10.4018/978-1-5225-5406-6.ch006

INTRODUCTION

The raising requirement for excessive data rates and then the limited obtainable bandwidth revokes the exploration of wireless devices that may effectively make use of the spatial domain (Subramanian & P., 2016). Considering the expense, overall size, and complexity constraints along at the terminal, antenna arrays are often taken into consideration merely along at the base stations (access points) to spatially discriminate the preferred signal via disturbance and noises. Making use of spatial diversity (Dietrich et al, 2001) at the same time on reception and transmission, can certainly make improvements to throughput and so coverage also to permitting a much higher level of spectral reuse and thus maximize the system capability. Therefore for higher data rate can be achieved with UWB and MIMO techniques. Antenna is essential component of any wireless system and hence in this chapter these two type of antennas to be analyze. The Chapter organized as follow: in the first steps the detailed literature of UWB and MIMO antenna have been carried out and in the next section the measurement analysis of monopole UWB and 2×2 MIMO antenna with monopole disc for UWB applications have been carried out. Finally chapter conclude with future direction and scopes.

UWB ANTENNA'S DESIRABLE PROPERTIES

The ultra-wide frequency bandwidth is the key parameter in differentiating the UWB antenna with various other antennas. In accordance with the FCC's specific description (Fields & R. E, 1997) an appropriate UWB antenna should have the ability to deliver an overall bandwidth of in no way lower than 600 MHz frequency or simply a fractional-bandwidth concerning at minimum 0.20 dB. In addition, UWB antenna should be remain working or operational. Also it must show steady impedance matching covering the whole frequency spectrum of 3.1-10.6 GHz in the circumstance of Impulsive-UWB in accordance with the outlined spectral mask defined in the FCC guidelines. In some cases, it might be needed (such as in European countries) in which UWB antennas need to be furnished the band-rejected feature with various other narrowband systems and applications accounting for the exact same operational band.

Depending on the nature of the application being processed, the directional or simply Omni-directional antenna characteristics are often desired in accordance to the functional software programs (Cordeiro et al., 2005) Omni-directional patterns are more often than not preferred in cellular and hand-held devices. Directional radiation patterns are generally preferred for radar units and additional directional unit in which high-gain is often required.

Substantial radiation efficiency is generally needed for antennas as far as it is essential and critical for an UWB antenna considering that the power spectral density being transmitted is overly low (Porcino et al., 2003). As a result, any kind of substantial loss received by the antenna may sometimes compromise the operation of the device.

An acceptable antenna requires to be compact as well low in weight for its suitability to various applications (Dietrich, Dietze, Nealy, & Stutzman, 2001). When we design UWB antenna for some specific requirements such as mobile and other portable (Mazhar et al., 2013) systems, it is generally advisable to design the UWB antenna into low profile feature and therefore it should also meet the compatibility condition for integration on printed base.

An effective structure involved with UWB antenna needs to remain optimum to get relatively high performance of the entire system. In order to avoid the likely in-band/out-band disturbance between the UWB devices and accessible other digital devices/instruments, the antenna must be designed in such a way that the entire system (antenna being used and the RF front-end system) is in accordance with the necessary power emission mask mentioned in by FCC and also various other regulatory organizations guidelines. The emission boundaries shall be driven by together with the selection of source pulse as well as the design of the UWB models/systems antennas.

The concept of UWB emerges from the time dependent behavior of electromagnetics, as a result UWB antenna is expected to show decent characteristics in the time-domain (e.g. some nice impulse response etc.). The assumption is merely which can define the Linear Time Invariant (LTI) unit using their response impulse in lieu of amplitude and phase measuring vs. radio frequency which is in general called the swept-frequency response (Dia, GODORA, Alicalapa, & Fabre, 2005). Pertaining to the narrow-band circumstance, it can be estimated that an antenna offers comparable functionality covering the whole bandwidth and the fundamental parameters, including gain, return loss, and possess small deviations throughout the operational band. On the other hand, Impulsive-UWB antenna generally utilize quite short pulses for data transmitting. Put another way, extensive bandwidth is often populated, hence the antenna won't be able to be viewed as a spot filter any longer but it will behave like band-pass filter. In such a circumstance, the antenna reveals additional sizeable influences on the input information. Consequently, a decent time domain operation (i.e., minimal pulse warping from the available signal) is a key challenge in the designing of a proper UWB antenna due to the fact that the signal is he carrier of very important details (Rappaport et al., 2002). As a result, it is essential and crucial to analyze the properties of the antenna under operation in the time domain environment.

Group delay is another significant variable to be carefully investigated (Chen et al., 2004) which usually signifies the degree of warping of the signal from the UWB antenna. Group delay can be described as a way of measuring the slope in the transmitted phased response. The linear segment of this phase response can be transformed into a prolonged amount thereafter the change by linear phase are converted straight to deviations caused by constant group delay. The changes during the group delay induce signal warping, in the same way deviations coming from linear phase induce warping

The prospects of UWB concept is substantial due to their remarkable positive aspects including the ability to allow excessive data rates at short transmission ranges due to their low power dissipation nature. The immediate progress in wireless communication networks has turned UWB an exceptional concept to displace the classic wireless systems in contemporary use just like Bluetooth as well as, wireless LANs, and many others. Numerous groundwork appears to have been in developing the UWB LNAs, mixers and so entire front-ends although not really much in developing the UWB antennas (Werbach & K, 2002). Most recently, educational and so commercial organizations have come to the realization that the trade-offs in between antenna developments and as well, transceiver complexity (Ghasemi et al., 2008). Generally, the transceiver complexity has become increased, together with the arrival of high level wireless transmitting approaches. To be able to improve the efficiency of transceiver while not compromising the expensive design, improved antenna structure needs to be utilized considering that the antenna is an essential element of the transceiver. Likewise, the complexity of this all round transceiver is lowered. To put into practice UWB technology, several concerns to conquer. UWB has a sizeable impact on antenna design. It has fascinated a upraise of curiosity in antenna design by means of cutting edge issues and prospects for antenna professional designers as UWB devices need to have an antenna with the operating bandwidth covering up the whole UWB (3.1- 10.6 GHz) and so in a position of receiving on companion frequencies along the precise same time frame. For that reason, the antenna patterns and functionality need to be reliable and consistent over the UWB. In addition, UWB is a technology which usually modulates impulse depending waveforms instead of continuous carrier waves. Consequently, the structure related with UWB antennas needs unique criteria from all those employed in developing narrowband antennas. The toughest dilemma in developing a UWB antenna is to accomplish vast impedance bandwidth that has substantial radiation efficiency (Alexiou et al., 2004). UWB antennas accomplish a bandwidth, in excess of 100% to the center frequency to ascertain sufficient impedance with the intention that merely only 10% of event signal can be dropped caused by reflections on the antenna's input port. A return loss of in excess of 10 dB is normally crucial to be able to attain substantial radiation potency. It is actually demanded as UWB transmission features rather low power (underneath the typical noise level) as well as, with substantial sensitivity.

There exists a growing requirements pertaining to compact and low cost UWB antennas which might be capable to provide adequate functioning in terms of both time as well as, frequency domains (Soh et al., 2015). The trend current wireless systems, along with UWB based devices, are to make modest, low-profile built- in circuits to be able to be appropriate together with the portable wireless systems. Likewise, the size of the antenna implicates the gain and so bandwidth. Subsequently, the size ın the antenna is taken into consideration among the severe challenges in UWB system. The employment of a planar design is able to diminish the volume of the UWB antennas just by exchanging 3-D radiators with their planar alternatives (Chen et al., 2015) On top of that, the 2-D geometry enables the fabrication less complicated. By doing this, the planar antenna is often imprinted over a suitable PCB and then swiftly integrated straight down into the RF circuit board. In recent years, a demand has been observed to raise the data rate among existing wireless communication devices. In this regard, the use of diversity approaches, most often accepting couple of antennas in a mobile terminal, can easily improve the data rate as well as, reliability without compromising spectrum or transmitted power in dense scattering conditions.

MULTIPLE-INPUT MULTIPLE-OUTPUT

Multiple-input-multiple-output (MIMO) technology has fascinated curiosity in contemporary wireless communication networks. MIMO devices transfer precisely the same power by using multiple antennas from the transmitter to the receiver in that way strengthening the channel capacity without need of added bandwidth or simply power. MIMO UWB devices can potentially maximize the channel capabilities in comparison to classic MIMO devices for narrowband practices. To deal with the multipath fading challenge within an indoor UWB wireless systems, an UWB diversity antenna strategy is definitely an encouraging choice. Having said that, for an effective MIMO antenna system mutual coupling amongst the antennas needs to below as far as feasible (Kildal et al., 2004). Therefore, all these structure challenges and so features intended for attaining substantial channel capacity by having significantly less complexity sparks the interest and so serves as an encouragement to the researchers in the research and design of MIMO antennas for substantial data rate UWB functions.

Background

Ultra-wideband (UWB) technology is absolutely not really new approach, even though it might have a very good cutting edge concept in designing the wireless communication systems. The idea of ultra -wideband technology commenced at the beginning of 1960's, whereby the exploration in time-domain electromagnetics be responsible for the application of impulse measurement approaches in the designing of wideband antennas (Lehpamer et al., 2016). This opened the way to the creation of short pulse radar devices. UWB was sometimes known to as 'baseband' or 'impulse' concept. Afterwards, with the aid of advances through this technology, a large number of approaches and execution strategies have been established for a number of applications which include radar, positioning systems and so short distance indoor applications, and many others.

MIMO is one of the most recent versions of intelligent antenna concept which can raise communication general performance (Jensen et al., 2004). The idea of spatial multiplexing implementing MIMO was initially unveiled in 1993. In the industrial vicinity, one of the primary device was designed in 2001, in which MIMO was utilized with the orthogonal frequency-division multiple access (MIMO-OFDMA), which in turn backed each of those diversity coding as well as, spatial multiplexing.

The advantages of MIMO concept turned out to be some of the most effective approaches to improve the channel capacity amongst the readily available bandwidth and so power. Most recently, micro strip antenna also implemented in MIMO technology by using a couple of radiating patches in the structure intended for transmission. Prior to now, quite a lot of MIMO/ Diversity antennas have already been recommended by various researchers which are now able to functioning in the frequency spectrum of 3.1 to 10.6 GHz, appropriate to UWB criteria. A couple of research have been accomplished on different MIMO antenna devices (Larsson et al., 2014) that has two and four radiating components and then numerous strategies have been suggested to raise the isolation between the antenna components.

The two most important applications of UWB antennas are merely lie wireless personal area network (WPAN) and other is Wireless body area network (WBAN) under the indoor environments where the dense multipath propagation becomes the principle causes of Inter Symbol Interference (ISI) (Ahmadi-Shokouh et al., 2009). Consequently, within order to transform this setback back in some positive aspects, multiple antennas or MIMO approaches is often utilized to take advantage of these kinds of rich scattering surroundings. Here, the fact of greater importance is that the UWB devices are quite often limited within the short distance communication because of really low transmission power allowed in the FCC guidelines. In this context, implementation of MIMO along with UWB allows advancing the communication range with some improvement in the link reliability (Sibille et al., 2010). The extensive many advantages of UWB-MIMO is usually summarized as subsequent; disturbance mitigation/suppression, excessive data-rates, superior link quality, prolonged exposure, simplified analogy components requisites, and even more the concurrent localization. In addition to all these added benefits, there are likewise issues in joint execution of UWB and MIMO. Such issues normally include: signaling trade-offs, channel modelling, various modulation optimization schemes, designing of modest and appropriate UWB antenna arrays, effective and affordable RF circuit layout, and many others. Amongst the many of such issues, the design of modest and appropriate UWB diversity antennas has become a huge hit by a large number of researchers. The importance of antenna with regards to wireless communication platform simply cannot be overlooked. This also turns into an essential aspect to always be miniaturized besides of several other components involved in the circuit. The strategies involved with antenna designing encounters a lot of challenges in itself. Seeing that, the systems are heading to be considerably more miniature, as a result the antennas ought to be positioned within an existing space. In addition, MIMO could be executed by three approaches; namely the spatial-multiplexing, beam-forming, and the diversity methods. Diversity methods, much precisely known as antenna-diversity methods (such as polarization, spatial, and the pattern) (Vaughan et al., 1987), are actually utilized during the MIMO antenna designs. Therefore, the rendering of multi-antenna architectures turns into even more difficult in the rather small space offered through compact terminals. The restrictions to take advantage of the diversity happen once the antennas are positioned in close neighborhood. Hence, it is usually advisable to de-correlate their particular patterns, or simply specifically, mutual coupling must get reduced (Khaleel et al., 2011). A further difficult task is an improvement of isolation ın between the working terminals of MIMO systems. In recent year, few MIMO antenna designs can be found in literature intended for UWB devices as follow:

In the (Najam et al.,2010) two antenna elements placed orthogonally to each other and isolation of $|S_{21}| \leq -15$ dB have been obtained with frequency range of 3.1 GHz-10.6 GHz. An UWB-MIMO antenna designed in (Li & J. F, 2012) using two symmetric monopole and two novel bent slits. Antenna offered good isolation of $|S_{21}| \leq -18$ dB from 2.4 -6.55 GHz only. Another monopole UWB-MIMO antenna with diversity performance fabricated for Wi-Fi/Wi-Max bands with isolation of $|S_{21}| \leq -17$ dB (See et al., 2012). A compact size antenna with 25×40 mm^2 size for UWB dongle applications is presented in (Zhang et al., 2012), and very good isolation of 26 dB obtained for lower UWB band 3.1 -5.15 GHz. An UWB- MIMO antenna with forked shape ground slit developed in (Prasanna et al., 2013) which offer the isolation of 20 dB for the frequency range of 4.4 - 10.7 GHz.

MAIN FOCUS OF THE CHAPTER

In this chapter two cases of measurement analysis is considered in the first case monopole Ultra-wide band antenna with defected ground structure is designed and measure in terms of UWB performance, and in the next case that antenna modified into a MIMO antenna. UWB performance of both antennas are compared in terms of return loss and group delay while other parameters of MIMO antenna to be discuss separately. Further UWB-MIMO performance of modified MIMO antenna compared with as above mention literature. Entire measurement analysis carried out using CST Microwave Simulation tool.

Monopole UWB Antenna With Defected Ground Structure

A circular monopole disc embedded on the FR4 substrate with relative permittivity of 4.4 and height of 1.59 mm. The overall size of monopole UWB antenna is 28.3 x 24 mm^2. A novel L shape slot etched in the partial ground plane of 12×24 mm^2. The defected ground slot offers good resonance match for the frequency range of 3.4 – 20 GHz .The radiating monopole disc element fed by matched 50 ohm micro strip line. The designed prototype is illustrated in Figure 1.

Figure 1. Structure of monopole UWB antenna with defected ground structure, L_{SUB} =28.3, W_{SUB} = 24, L_G =12, W_G = 24, L_{C1} = 9, L_{C2} = 3 (all dimensions are in mm)

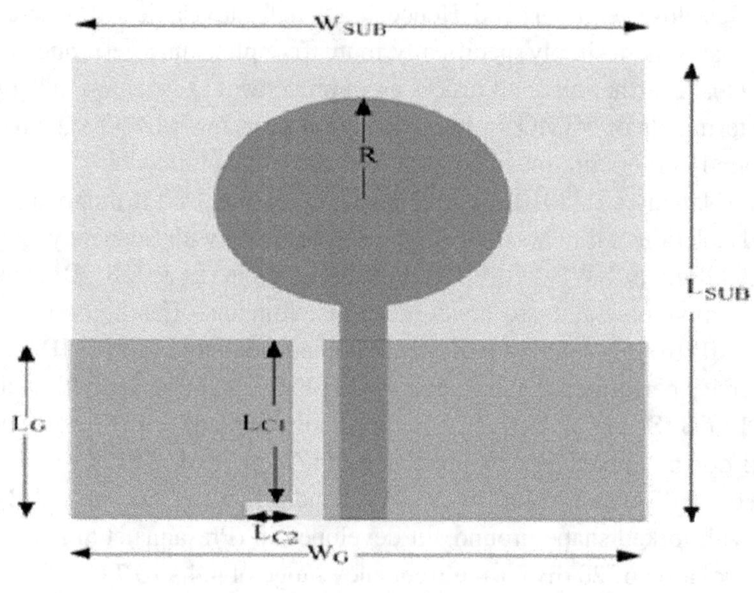

Results of return loss of proposed monopole UWB antenna with defected ground structure is depicted in Figure 2. The result indicates here impedance band width 3.9-20 GHz and hence entire frequency as per UWB band (3.1-10.6 GHz) has been covered by monopole UWB antenna. Further this antenna modified and converted into MIMO antenna for UWB applications as describe below:

2×2 MIMO ANTENNA WITH MONOPOLE DISC FOR UWB APPLICATIONS

The proposed MIMO antenna with two monopole disc on the FR4 substrate with small size of 64×40 mm^2, thickness and relative permittivity of 0.8 mm, 4.4 respectively is designed. On the opposite side a partial ground plane is formed with the size of 11.5×64 mm^2. These two circular monopole discs are 34 mm apart from each other and their radius are 11.5 mm. The novel ground slits embed in the ground plane which offers good isolation of 18 dB. The proposed MIMO antenna is depicted in Figure 3.

Figure 2. Result of return loss of monopole UWB antenna

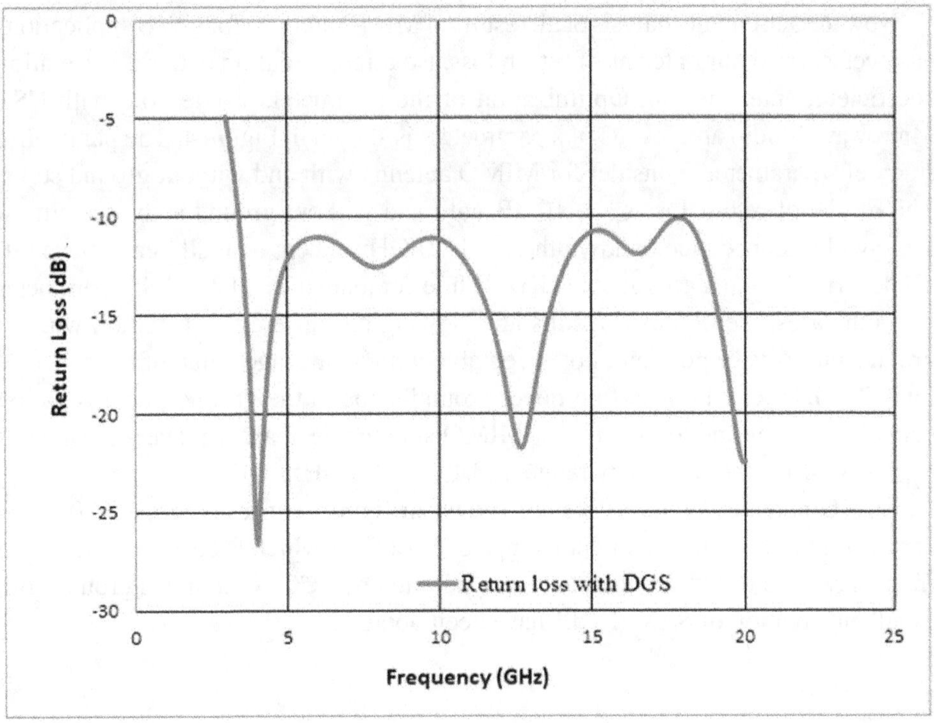

Figure 3. Structure of MIMO antenna with two monopole disc and inserted ground stubs

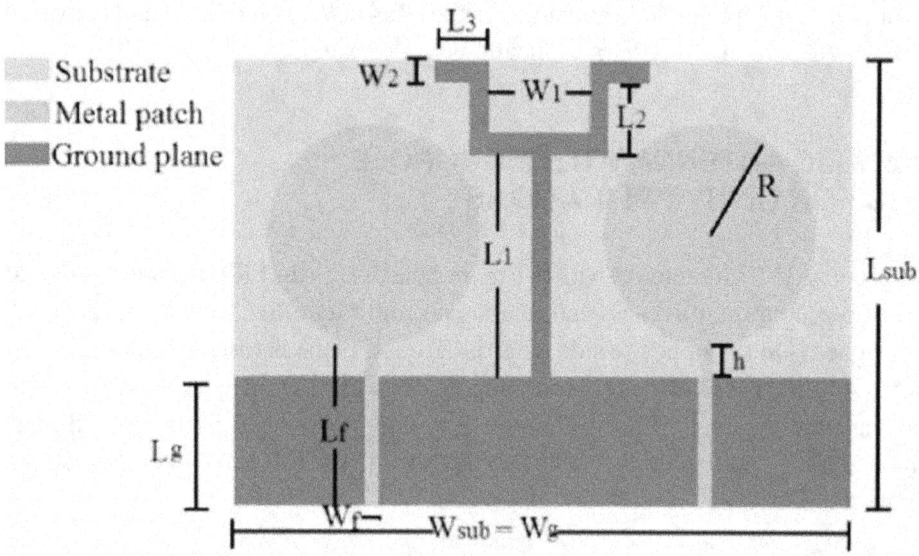

Now measurement analysis of all results of MIMO antenna for UWB applications has been carried out in terms of return loss, isolation, radiation pattern, correlation coefficient, antenna gain. Optimization of the parameters carried out with CST Microwave Studio and results of s parameters is shown in Figure 4. The parametric study of s parameters consider for MIMO antenna with and without ground stubs. The results of return loss s_{11} <-10 dB with and without ground stubs are almost same with a impedance bandwidth of 3.1-20 GHz except a small peak above -10 dB occurred in both cases at 15.2 GHz. While compare the isolation (S_{21}) parameter for both cases, the obtained results are showing big difference. Isolation without ground stub is very poor and not acceptable with an average value of S_{21} <-10 dB for 3-20 GHz .On the inserting novel ground stubs in the ground plane isolation between the both monopole disc of MIMO antenna decreases an average value of $S_{21} \leq$ -18 dB for the frequency range of 3.1 GHz-20 GHz.

MIMO antenna s parameters are shown in figure return loss S_{11} < -10 dB is very good for the entire frequency range of 3-12 GHz which is cover the interested frequency band of Ultra-wideband freely allotted by FCC. With novel ground stub excellent isolation of $S_{21} \leq$ -18 dB have been obtained.

Figure 4. S parameters results of MIMO antenna

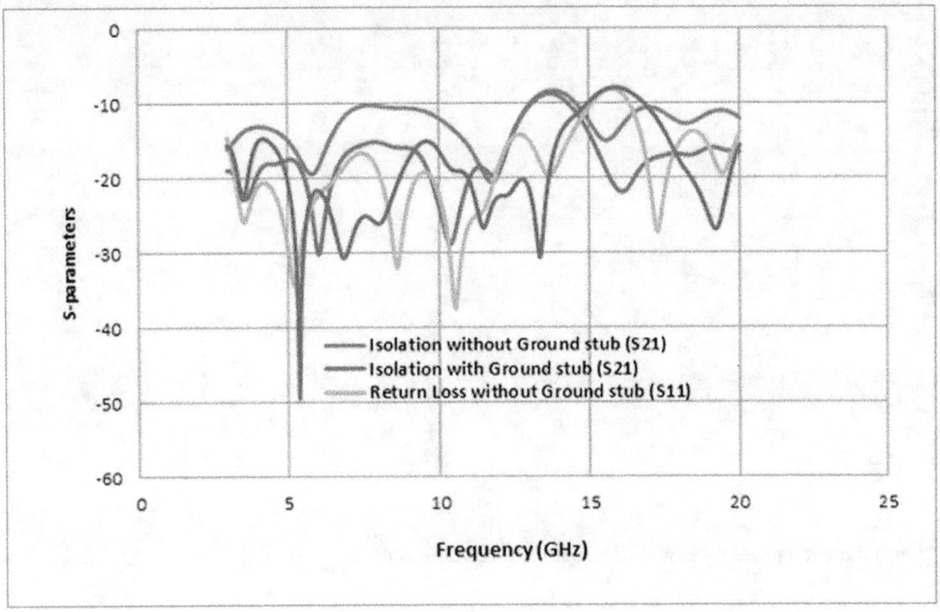

**For a more accurate representation see the electronic version.*

The efficiency of MIMO antenna is hard to stick to mutual coupling among the multiple antennas. To achieve low mutual coupling between such antenna elements, three distinct type of diversity techniques (spatial, polarization and pattern diversity) are being used in common practice. Nevertheless, spatial diversity technique is not necessarily being convenient when considering size and as well, cost of MIMO antenna. With regard to our case pattern diversity techniques is followed which make feasible to design compact size of MIMO diversity antenna without degrade its performance

The radiation pattern of proposed MIMO antenna accomplished utilizing CST microwave studio and results are studied at 5, 6 and 8 GHz frequencies, when only one circular monopole is excited while other circular monopole is terminated with matched 50 Ω load. Results are as depicted in Figure 5. It has been noticed that the radiated power density from circular monopole 1 in positive X direction and from circular monopole 2 in negative X direction so because of this patterns are just opposite to each other well, and consequently radiation patterns altogether follow the pattern diversity to reduce the mutual coupling dilemma between the two antenna elements in multipath fading environment.

Figure 5. Results of 3D radiation patterns at (a) 5GHz frequency, (b) 6 GHz frequency, (c) 8 GHz frequency

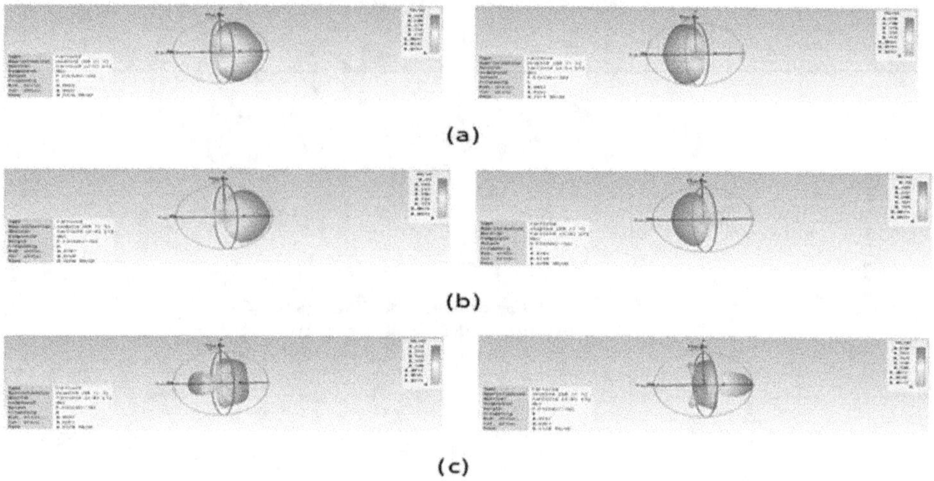

**For a more accurate representation see the electronic version.*

Correlation coefficient can be described as parameter that could be considerably more vital for MIMO systems. The trans-receiver signal from or to the antenna might be correlated to some extent. It's actually a static parameter that measures the level of similarity among the received signal, its modulus value varies from 0 to 1 and preferably it ought to be zero or tolerable value below 0.5 as it possibly can for good diversity performance of MIMO systems. This parameter can be calculated by both equally 3D radiations pattern or by using S parameter. The correlation coefficient of proposed antenna assessed working with S parameter and it is observed in figure 6 that their value is very low i.e. down below .05 with in interesting band of frequency (3.1 GHz-10.6 GHz) except for a peak arises at the 6.86 GHz frequency though this value is likewise really low i.e. 0.10. We deduce right from these results that the proposed antenna offers good diversity performance and high isolation in terms correlation coefficient.

The obtained peak gain of MIMO antenna is shown in Figure 7. it can be brief that peak gain is varying from 3.2 dB to a maximum value of 4.6 dB within the ultra-wideband (3.2-20 GHz) with a peak value of gain 7.04 dB at the frequency of 12.59 GHz.

The diversity gain is a quantity accustomed to define the diversity overall performance of MIMO antenna. The diversity gain is often calculated in diverse way which include apparent diversity gain, effective diversity gain and actual diversity gain. Each one of these gains determined with the help of difference between power

Figure 6. Results of Correlation coefficient of MIMO antenna

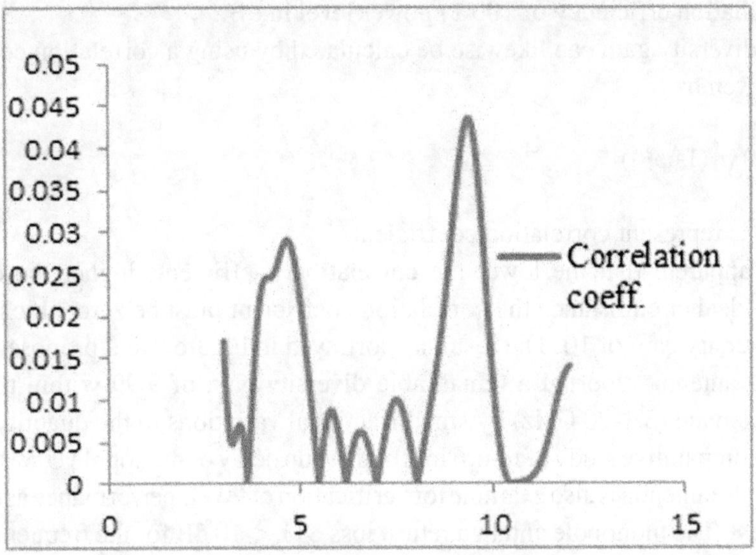

Figure 7. Gain of MIMO antenna

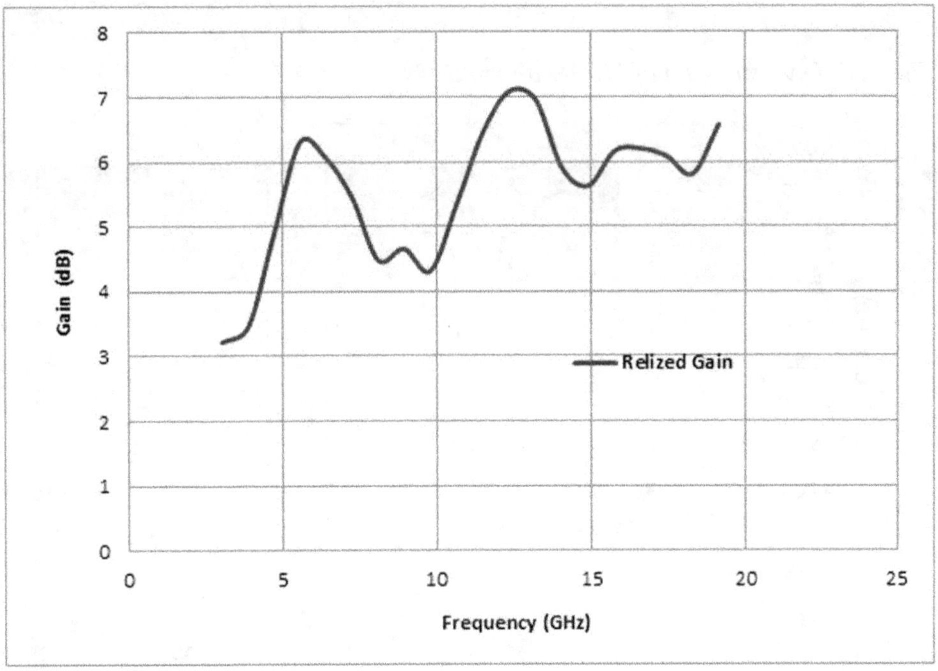

levels in dB of combined signal of practical MIMO antenna and reference antenna (with radiation efficiency of 100%) power level in dB.

The diversity gain can likewise be calculated by using a correlation coefficient and is given by:

$$DG = 10\sqrt{(1-\backslash\rho^2 \backslash)} \quad (1)$$

where ρ is represent correlation coefficient

It is apparent that the lower the correlation coefficients higher the diversity gain .In ideal circumstance the correlation coefficient must be zero which provide ideal diversity gain of 10. The result as portrayed in Figure 8 and proposed MIMO diversity antenna reported a remarkable diversity gain of 9.99 within the entire ultra-wideband (3.1-20 GHz) by small fractional variations in the quantity.

The comparative study of return loss and group delay of monopole UWB antenna with MIMO antenna is also examine for verification of UWB performance as depicted in figure 8. The monopole antenna return loss S11 <-10 dB for the frequency range of 3.9-20 GHz whereas MIMO antenna covers the lower frequency of UWB with a frequency range of 3.1-20 GHz except a peak above -10 dB occurred at frequency range of 14.9-16.8 GHz. The Group delay of both antennas is almost constant and same except a peak occurred in MIMO antenna at 5.3 GHz with a value of -1.42.

Figure 8. Diversity gain of MIMO antenna

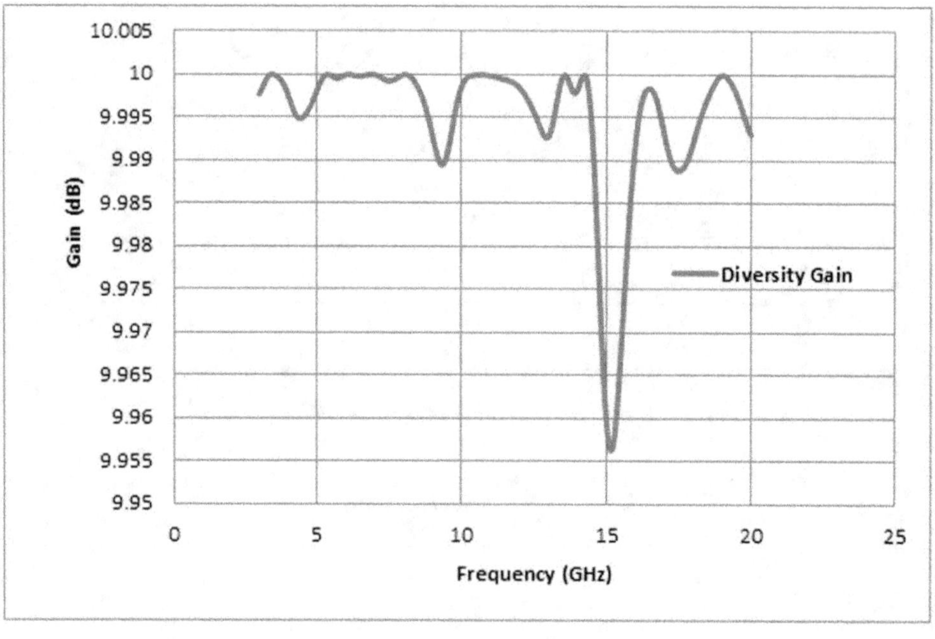

Figure 9. Comparative analysis of return loss of monopole UWB antenna with MIMO antenna

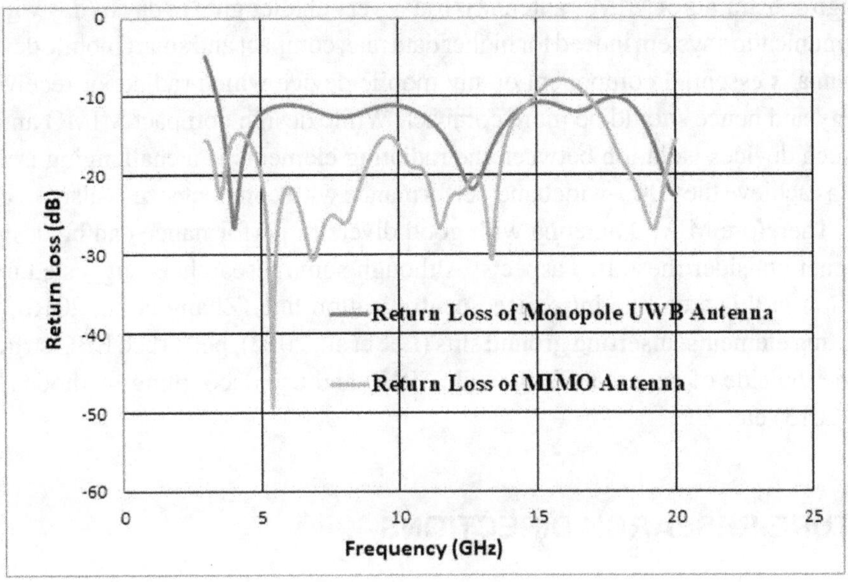

**For a more accurate representation see the electronic version.*

Figure 10. Comparative analysis of group delay of monopole UWB antenna with MIMO antenna

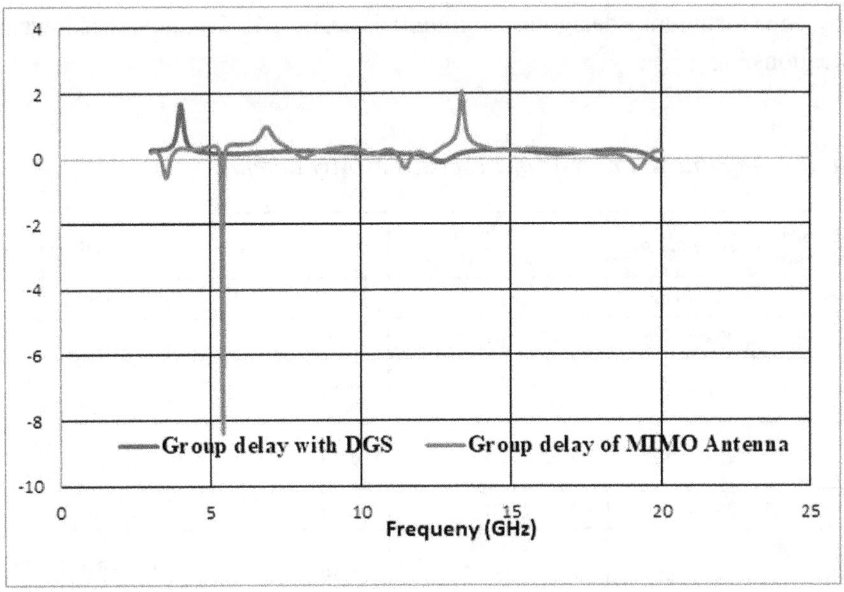

**For a more accurate representation see the electronic version.*

SOLUTIONS AND RECOMMENDATIONS

Design constraints of MIMO antenna is today very challenging as day by day wireless communication system indeed for higher data rate, compact and smart mobile devices. Antenna is essential component of any mobile device which radiate or receive RF energy and hence should be more compact. While design compact MIMO antenna for such devices isolation between the radiating elements is a challenging issue as well as achieve the Ultra-wideband performance with compact size is also a tedious task. Therefore MIMO antenna with good diversity performance can be design if designer consider these two aspects. Although some researchers suggested useful solution in this area like introduced neutralization line (Zhang et al., 2016), with radiating elements, inserting ground slits (Lee et al., 2012), perturbed EBG structure on the top side of antenna (Yang et al., 2017) and use decoupling methods (Li & J. F, 2013) etc.

FUTURE RESEARCH DIRECTIONS

As demanding of higher data rate, wireless technology will be shift towards Massive MIMO in 5G technology. Therefore design of MIMO antenna for numerous applications is emerging trend in the new era of wireless technology. Internet of thing in this era is very useful for wearable devices for medical applications. As already discuss antenna play important role in wireless systems therefore another future scope for antenna designer community is design MIMO antennas for wearable applications.

Table 1. Comparative analysis of Proposed MIMO antenna

Antenna Reference	Antenna size	Impedance Bandwidth	Isolation	Radiation Pattern
Najam et al.	65× 40 mm^2	2.4-6.8 GHz	-15 dB	Directional
Li & J. F	40×90 mm^2	2.4-4.2 GHz	-17 dB	Directional
See et al.,	78×40 mm^2	2.4-6.55 GHz(lower uwb band)	-18 dB	Directional with pattern diversity
Zhang et al	25×40 mm^2	3.1-5.15 GHz	-26 dB	Directional with pattern diversity
Prasanna et al	35×40 mm^2	4.4-10.7 GHz	-20 dB	Omini directional
Proposed MIMO Antenna	**64×40 mm^2**	**3.1-20 GHz**	**-18 dB**	**Directional with pattern diversity**

CONCLUSION

UWB concept joined together with the allot more transmit and so receive antennas is a worthwhile approach in order to really attain data rates at ın excess of 1 Gb/s intended for wireless communication structure. UWB is more often than not utilized in short-range and so primarily indoor communications conditions portrayed commonly by dense multipath propagation. Involving this category of setting, MIMO devices enable to get a considerable rise in spectral performance by taking advantage of the built in array gain and so spatial multiplexing gain in the devices. In this chapter, we present a comprehensive analysis for UWB-MIMO wireless engineering, technological innovation.

This chapter also includes important aspects of Monopole UWB antenna design and MIMO antenna design for UWB applications. Comparative analysis of their impedance bandwidth and group delay has been carried out to verify their UWB performance. The UWB performance of MIMO antenna is better than monopole UWB antenna and other parameters like pattern diversity, correlation coefficient, and impedance bandwidth of MIMO antenna are appreciable. Therefore, this MIMO antenna may be a good choice for next generation wireless applications.

ACKNOWLEDGMENT

I am very thankful to Dr. Satya P. Singh who encourage us to elaborate about latest technology of MIMO antenna design for UWB application.

REFERENCES

Ahmadi-Shokouh, J., & Qiu, R. C. (2009). Ultra-wideband (UWB) communications channel measurements–a tutorial review. *International Journal of Ultra Wideband Communications and Systems, 1*(1), 11–31. doi:10.1504/IJUWBCS.2009.026447

Alexiou, A., & Haardt, M. (2004). Smart antenna technologies for future wireless systems: Trends and challenges. *IEEE Communications Magazine, 42*(9), 90–97. doi:10.1109/MCOM.2004.1336725

Chen, X., Zhang, Z., Chen, H. H., & Zhang, H. (2015). Enhancing wireless information and power transfer by exploiting multi-antenna techniques. *IEEE Communications Magazine, 53*(4), 133–141. doi:10.1109/MCOM.2015.7081086

Chen, Z. N., Wu, X. H., Li, H. F., Yang, N., & Chia, M. Y. W. (2004). Considerations for source pulses and antennas in UWB radio systems. *IEEE Transactions on Antennas and Propagation, 52*(7), 1739–1748. doi:10.1109/TAP.2004.831405

Cordeiro, C., Challapali, K., Birru, D., & Shankar, S. (2005, November). IEEE 802.22: the first worldwide wireless standard based on cognitive radios. In *New Frontiers in Dynamic Spectrum Access Networks, 2005. DySPAN 2005. 2005 First IEEE International Symposium on* (pp. 328-337). IEEE.

Dia, S., Godora, B., Alicalapa, F., & Fabre, A. (2005). Ultra wide-band: State of the art; implementation of a performance-controllable low-noise amplifier. *Turkish Journal of Electrical Engineering and Computer Sciences, 13*(1), 1–23.

Dietrich, C. B., Dietze, K., Nealy, J. R., & Stutzman, W. L. (2001). Spatial, polarization, and pattern diversity for wireless handheld terminals. *IEEE Transactions on Antennas and Propagation, 49*(9), 1271–1281. doi:10.1109/8.947018

Fields, R. E. (1997). Evaluating compliance with FCC guidelines for human exposure to radiofrequency electromagnetic fields. *OET Bulletin, 65*, 10.

Ghasemi, A., & Sousa, E. S. (2008). Spectrum sensing in cognitive radio networks: Requirements, challenges and design trade-offs. *IEEE Communications Magazine*, *46*(4), 32–39. doi:10.1109/MCOM.2008.4481338

Jensen, M. A., & Wallace, J. W. (2004). A review of antennas and propagation for MIMO wireless communications. *IEEE Transactions on Antennas and Propagation*, *52*(11), 2810–2824. doi:10.1109/TAP.2004.835272

Khaleel, H. R., Al-Rizzo, H. M., Rucker, D. G., Rahmatallah, Y. A., & Mohan, S. (2011, July). Mutual coupling reduction of dual-band printed monopoles using MNG metamaterial. In *Antennas and Propagation (APSURSI), 2011 IEEE International Symposium on* (pp. 2219-2222). IEEE. 10.1109/APS.2011.5996956

Kildal, P. S., & Rosengren, K. (2004). Correlation and capacity of MIMO systems and mutual coupling, radiation efficiency, and diversity gain of their antennas: Simulations and measurements in a reverberation chamber. *IEEE Communications Magazine*, *42*(12), 104–112. doi:10.1109/MCOM.2004.1367562

Larsson, E. G., Edfors, O., Tufvesson, F., & Marzetta, T. L. (2014). Massive MIMO for next generation wireless systems. *IEEE Communications Magazine*, *52*(2), 186–195. doi:10.1109/MCOM.2014.6736761

Lee, J. M., Kim, K. B., Ryu, H. K., & Woo, J. M. (2012). A compact ultrawideband MIMO antenna with WLAN band-rejected operation for mobile devices. *IEEE Antennas and Wireless Propagation Letters*, *11*, 990–993. doi:10.1109/LAWP.2012.2214431

Lehpamer, H. (2016). *Introduction to power utility communications*. Artech House.

Li, J. F., Chu, Q. X., & Huang, T. G. (2012). A compact wideband MIMO antenna with two novel bent slits. *IEEE Transactions on Antennas and Propagation*, *60*(2), 482–489. doi:10.1109/TAP.2011.2173452

Li, J. F., Chu, Q. X., Li, Z. H., & Xia, X. X. (2013). Compact dual band-notched UWB MIMO antenna with high isolation. *IEEE Transactions on Antennas and Propagation*, *61*(9), 4759–4766. doi:10.1109/TAP.2013.2267653

Mazhar, W., Tarar, M. A., Tahir, F. A., Ullah, S., & Bhatti, F. A. (2013). Compact microstrip patch antenna for ultra-wideband applications. *PIERS Proceedings*, 1100-1104.

Najam, A. I., Duroc, Y., & Tedjini, S. (2010, April). Design and analysis of MIMO antennas for UWB communications. In *Antennas and Propagation (EuCAP), 2010 Proceedings of the Fourth European Conference on* (pp. 1-5). IEEE.

Porcino, D., & Hirt, W. (2003). Ultra-wideband radio technology: Potential and challenges ahead. *IEEE Communications Magazine*, *41*(7), 66–74. doi:10.1109/MCOM.2003.1215641

Prasanna, K. M., & Behera, S. K. (2013, April). Compact two-port UWB MIMO antenna system with high isolation using a fork-shaped structure. In *Communications and Signal Processing (ICCSP), 2013 International Conference on* (pp. 726-729). IEEE. 10.1109/iccsp.2013.6577151

Rappaport, T. S., Annamalai, A., Buehrer, R. M., & Tranter, W. H. (2002). Wireless communications: Past events and a future perspective. *IEEE Communications Magazine*, *40*(5), 148–161. doi:10.1109/MCOM.2002.1006984

See, C. H., Abd-Alhameed, R. A., Abidin, Z. Z., McEwan, N. J., & Excell, P. S. (2012). Wideband printed MIMO/diversity monopole antenna for WiFi/WiMAX applications. *IEEE Transactions on Antennas and Propagation*, *60*(4), 2028–2035. doi:10.1109/TAP.2012.2186247

Sibille, A., Oestges, C., & Zanella, A. (2010). *MIMO: from theory to implementation*. Academic Press.

Soh, P. J., Vandenbosch, G. A., Mercuri, M., & Schreurs, D. M. P. (2015). Wearable wireless health monitoring: Current developments, challenges, and future trends. *IEEE Microwave Magazine*, *16*(4), 55–70. doi:10.1109/MMM.2015.2394021

Subramanian, P. (2016). *Improving reliability aware flooding algorithms in wireless multi-hop sensor networks*. Academic Press.

Vaughan, R. G., & Andersen, J. B. (1987). Antenna diversity in mobile communications. *IEEE Transactions on Vehicular Technology*, *36*(4), 149–172. doi:10.1109/T-VT.1987.24115

Werbach, K. (2002). The New Wireless Paradigm. *Spectrum Series Working Paper, 6*.

Yang, X., Liu, Y., Xu, Y. X., & Gong, S. X. (2017). Isolation Enhancement in Patch Antenna Array With Fractal UC-EBG Structure and Cross Slot. *IEEE Antennas and Wireless Propagation Letters*, *16*, 2175–2178. doi:10.1109/LAWP.2017.2703170

Zhang, S., Lau, B. K., Sunesson, A., & He, S. (2012). Closely-packed UWB MIMO/diversity antenna with different patterns and polarizations for USB dongle applications. *IEEE Transactions on Antennas and Propagation*, *60*(9), 4372–4380. doi:10.1109/TAP.2012.2207049

Zhang, S., & Pedersen, G. F. (2016). Mutual coupling reduction for UWB MIMO antennas with a wideband neutralization line. *IEEE Antennas and Wireless Propagation Letters*, *15*, 166–169. doi:10.1109/LAWP.2015.2435992

Chapter 7
Developments of Environmentally Certified Reference Material From the Brazilian Metrology Institute to Support National Traceability

Andreia de Lima Fioravante
National Institute of Metrology Standardization and Industrial Quality, Brazil

Evelyn de Freitas Guimarães
National Institute of Metrology Standardization and Industrial Quality, Brazil

Fabiano Barbieri Gonzaga
National Institute of Metrology Standardization and Industrial Quality, Brazil

Cristiane Rodrigues Augusto
National Institute of Metrology Standardization and Industrial Quality, Brazil

Claudia Cipriano Ribeiro
National Institute of Metrology Standardization and Industrial Quality, Brazil

Eliane Cristina Pires do Rego
National Institute of Metrology Standardization and Industrial Quality, Brazil

Elaine Batista de Santana
National Institute of Metrology Standardization and Industrial Quality, Brazil

Laura Alves das Neves
National Institute of Metrology Standardization and Industrial Quality, Brazil

Lucas Junqueira de Carvalho
National Institute of Metrology Standardization and Industrial Quality, Brazil

Renato Rubim Ribeiro de Almeida
National Institute of Metrology Standardization and Industrial Quality, Brazil

DOI: 10.4018/978-1-5225-5406-6.ch007

Rodrigo C. de Sena
National Institute of Metrology Standardization and Industrial Quality, Brazil

Janaina Marques Rodrigues Caixeiro
National Institute of Metrology Standardization and Industrial Quality, Brazil

Marcelo de Almeida Dominguez
National Institute of Metrology Standardization and Industrial Quality, Brazil

Valnei Smarçaro da Cunha
National Institute of Metrology Standardization and Industrial Quality, Brazil

Sidney P. Sobral
National Institute of Metrology Standardization and Industrial Quality, Brazil

ABSTRACT

This chapter aims to present the developments performed by the Brazilian Metrology Institute (NMI) – Inmetro, considering the environmental demand. Inmetro addresses a great part of its activities to the study of the traceability transference based on production and dissemination of certified reference material (CRM) of different areas in chemistry. The chapter presents results from certification of the following reference materials developed: BTEX and PAH in solution, besides automotive emission gas mixtures and bioethanol. So, the achievements made are the growth in developing CRM in order to support the needs of the national industry and to disseminate traceability among the society.

INTRODUCTION

The results from chemical measurements form the main basis for decision-making in economic, political, environmental, medical or legal fields. The ongoing economic globalization requires that these results are reliable, comparable and traceable. The Brazilian National Metrology Institute (NMI), Inmetro, is a locus of knowledge and credibility, committed to the diffusion of metrology culture among society, as well as representation of the Country as international articulation. The Chemical and Thermal Metrology Division from Inmetro (Dimqt) has 7 (07) laboratories: Organic, Inorganic, Electrochemistry, Gases, Physical Chemistry, Humidity and Thermal Laboratories. All these branches from Dimqt plays an indispensable role, mainly towards the development of certified reference material (CRM), focusing the necessity concerning trustworthy and accurate chemical analysis. Therefore, the importance of a NMI is to provide ways of supporting reliable measurements, in order to defend national interests, transpose technical barriers and ensuring fair terms to trade, supply harmonization and confidence to different sectors of society, aiming to develop products and /or services in the country with quality, innovation and competitiveness.

CERTIFIED REFERENCE MATERIALS

Certified reference materials (CRM) is one of the essential pillar of metrology in chemistry, as its use is an important tool in the implementation of various aspects of measurement quality and confidence. Another important observation is environmental legislation, that has significantly increased, and which requires international standards of comparison. These standards are essential to the realization of the SI units and, therefore, are indispensable for establishing traceability of measurements. Inmetro is mainly responsible for keeping the national metrological standards of reference, as well as for carrying and spread the SI units of measurement, and its harmonization on a global level, providing traceability to a large number of users via a metrological chain.

According to the International Organization for Standardization (ISO) definition (ISO Guide 30, 2008), a reference material (RM) is a material, sufficiently homogeneous and stable with respect to one or more specified properties, which has been established to be fit for its intended use in a measurement process. A RM can be either a pure substance or a mixture of components. Certified Reference Material (CRM) is a reference material characterized by a metrologically valid procedure for one or more specified properties, accompanied by a certificate that provides the value of the specified property, its associated uncertainty, and a statement of metrological traceability. One or more parameters of a CRM can be determined and confirmed according to a specific procedure. To each certified parameter an uncertainty should be assigned at a given confidence level. Certificates are issued by recognized bodies.

The production is performed according to the ISO 17034 (ABNT NBR ISO 17034, 2017) and ISO Guide 35 (ABNT ISO Guia 35, 2012) and includes studies of homogeneity, stability and characterization, and its respective estimation of uncertainties. The evaluation of the homogeneity is particularly important in cases where the substance to be measured is present in a complex matrix. The stability study is divided into short-term and long-term stability studies. The first simulates the conditions of transport and should address the factors involved at this stage that can affect the CRM, such as temperature and time. The results of this study show that the form of transport and packaging are appropriate to the material. In the study of long-term stability is developed to assess whether the material is stable during its period of validity conditions of storage. The variations observed during this period in the amount of the certified properties should be included in the determination of uncertainty. At the stage of characterization may be used the following methods: a primary method, two or more independent methods on a same laboratory, or by an interlaboratorial comparison using one or more validated methods of accuracy. These guides may provide the basis for future assessment of reference material producers.

So, the use of CRM ensures that the results obtained by different laboratories are comparable and traceable. In this context, CRM are essential in maintaining a system of universal and consistent measurements make it possible to compare the measured values from one laboratory to another, besides disseminating traceability among society.

ENVIRONMENTAL CRM FROM INMETRO

Automotive Emission Gas Mixtures

In the industrial and automotive sector, the growth and development of new technologies and the increasing number of cars on the roads, has derived a detrimental effect on the environment by changing the composition of the air and causing the warming of the earth's surface – the greenhouse effect – which is due to high concentration of certain gases that cause retention of infrared radiation in the atmosphere. A source of air pollution depends, first of all, to set benchmarks of the pollutants and their effects on humans, animals, plants or other materials, as well as the criteria to measure the frequency of occurrence of pollutants and their effects. Nowadays, vehicle emissions play an important role in the level of air pollution in cities, as represent approximately 40% of air pollution. The analysis of air quality by the responsible agencies requires reliable and comparable analytical methods in order to reduce the emission gases, such as carbon monoxide (CO) and carbon dioxide (CO_2), all of them relevant for implementing regulations with regard to car exhaust gas measurements.

This CRM developed by the Laboratory of Gas Analysis from Inmetro consists in a primary reference mixture (PRM) of carbon monoxide (CO), carbon dioxide (CO_2) and propane (C_3H_8) in nitrogen, gravimetically prepared. The bottling of the mixture occurs by a concentration demand of the client in individual aluminium cylinders of 5 L with an average pressure of 120 bar.

The production of these standards intends to avoid the importation needs and substantially decrease its costs, in order to include Brazil in the selected group of national producers of PRM, in order to support the demand of national gas industry.

Certification Procedures

The gravimetric preparation, as well as the calculation of the gas composition and associated uncertainty evaluation should be done in accordance with ISO 6142 (ISO 6142, 2015). This produced mixture was verified gas chromatograph (GC) equipped with thermal conductivity detector (TCD) for the CO and CO_2 analysis and a flame ionization detector (FID) for the propane analysis, which were performed in 03 (three) different days under repeatability conditions with 10 (ten) runs each day. Based on the calibration curve, analysis values and its respective uncertainties were assigned to the amount fractions of carbon monoxide, carbon dioxide, and propane using the ISO 6143 (ISO 6143, 2001).

This primary standard gas mixture (PSM) was submitted for the first participation of Inmetro in the international key comparison (KC) of PSM of automotive emission gravimetric production between known National Metrology Institutes (NMIs) under auspices of the Euramet (Van der Veen et al., 2011). This participation aimed to fulfil the necessary preliminary criteria for mutual recognition of measurements, and the results obtained from Inmetro were considered satisfactory.

A long-term stability study was performed during the period of 01 (one) year, with mensal analysis to compose the final estimation from this uncertainty contribution of stability of the mixture. It's important to observe that for gas mixture standards no homogeneous nor short-term stability study are necessary, as the batches are prepared individually by the client demand, besides considering the homogeneous physical-chemistry behaviour of those gas mixtures.

The certification of the gas mixture CRM is done after the stability study finishes, and the combined standard uncertainty (uCRM) is estimated according to Equation 1.

$$u_{cert} = \sqrt{u_{prep}^2 + u_{ver}^2 + u_{st}^2} \quad (1)$$

where:

u_{prep} = uncertainty of gravimetric preparation;
u_{ver} = uncertainty from verification (GC analysis);
u_{st} = uncertainty from stability.

The certified value presents the expanded uncertainty (U), and it is expressed as the product of the combined standard uncertainty (u_{CRM}) and the coverage factor k=2 (approximately 95% of confidence level) (JCGM, 2008; EURACHEM/CITAC, 2000).

Certification of Automotive Emission Gas Mixture in Nitrogen

According to studies observed for the CRM of automotive emission (CRM 8634), the cylinder containing the CRM should be handled with care and by experienced personnel in a laboratory equipped for the safe handling of gaseous materials. This CRM should be stored preferably at the temperature of 10 a 35 °C. A minimum pressure of 1 Mpa should be used. It's important to remark that Inmetro maintains a monitoring program of all CRM. Any change to the certified value observed during the monitoring will be immediately communicated to the user.

An example of a certified value and its expanded uncertainties from a PRM of automotive emission previously prepared for the gas industry (CRM 8634.0035) is presented in Table 1 (Inmetro, 2013a).

Table 1. Certified value and expanded uncertainty of CRM 8634.0035 from Inmetro (Brazil)

Compound in Nitrogen	Certified Value (mol/mol)
Carbon monoxide	(0.498 ± 0.004) x 10^{-2}
Carbon dioxide	(13.93 ± 0.07) x 10^{-2}
Propane	(196.74 ± 1.97) x 10^{-6}

The Gas Analysis Laboratory from Inmetro provides gas standards for calibrating vehicle emission monitors, which can be prepared to the requirements of the customer. The Table 2 gives an indication of the available mixtures presenting their range of amount in fraction (mol/mol) along with its specific expanded uncertainty, which are available on request.

Aromatic Organic Compounds (BTEX and PAH)

With the growing concern about the importance of a clean environment there is an urgent demand for certified reference materials of aromatic organic compounds, due to the high toxicity of these compounds. Several of these contaminants are listed as priority pollutants under the Environmental Protection Agency (US/EPA) and regulators in many countries.

Among the family of volatile organic compounds (VOC) it is highlighted the BTEX (benzene, toluene, ethylbenzene, o-xylene, m-xylene and p-xylene) in methanol, which may be carcinogenic and furthermore may have neurotoxic effects and/or mutagenic. The bottling of the solution occurs in individual amber ampoule with 1.5 mL/ampoule.

Another strand from aromatic organic compounds are the Polycyclic Aromatic Hydrocarbons (PAH) that are environmental pollutants found in water, air and soil, and considered of interest to public health due to environmental persistence and potentially toxic (mutagenic and carcinogenic) (IARC, 2010). This CRM consists in a solution of naphthalene, acenaphthylene, acenaphthene, fluorene, phenanthrene, anthracene, fluoranthene, pyrene, benzo[a]anthracene, chrysene, benzo[b]fluoranthene, benzo[k] fluoranthene, benzo[a]pyrene, indeno[123-cd]pyrene, dibenzo[a,h]anthacene and benzo[ghi]perylene in toluene, gravimetically prepared. The bottling of the solution occurs in individual amber ampoule with 1.2 mL/ampoule.

Table 2. Range of automotive emission in nitrogen CRM from Inmetro (Brazil)

Compound	Certified Value (mol/mol)	Relative Expanded Uncertainty (U) (%)[a]
Carbon monoxide	(10 - 1000) x 10^{-6}	0.5 – 0.2
Carbon monoxide	(1 - 100) x 10^{-3}	0.75 – 0.6
Carbon dioxide	(10 - 200) x 10^{-3}	0.5 – 0.2
Carbon dioxide	(1 - 100) x 10^{-3}	0.75 – 0.5
Propane	(300 - 50000) x 10^{-6}	1.0 – 0.6

[a]With coverage factor k=2 (approximately 95% of confidence level)

For this reason, the analysis of these environmental contaminants with a metrological focus is considered essential, which main goal is to provide quality and reliability for the measurements with traceability by the use of CRM of such compounds.

Certification Procedures

For the certification of the two CRM (BTEX and PAH in solution), gravimetric preparation method and isotope dilution mass spectrometry (IDMS) were used. These techniques are considered primary methods by the Consultative Committee for Amount of Substance (CCQM) (Milton & Quinn, 2001). Before the certification studies, methods by gas chromatography/ isotope dilution mass spectrometry (GC-MS/IDMS) were optimized and validated for the determination of BTEX and PAH in solution. All performance parameters evaluated were appropriate to the scope of the study. These developed methods showed good measurement accuracy and precision, indicating the efficiency of the quantification method used.

The gravimetric concentrations of all standards were adjusted for the purity estimation of each component, which was either determined by using flame ionization capillary gas chromatography (GC-FID) with two stationary phases of different polarities and differential scanning calorimetry (DSC) for BTEX standards, and by GC-FID with two stationary phases of different polarities and quantitative nuclear magnetic resonance (qNMR) for PAH standards (Guimarães et al., 2015).

The characterization uncertainty includes all sources of measurement uncertainty, including the purity of the standards used in the methodology. The purity was determined by Inmetro following the NIST purity Guide (Duewer, Parris, White, May & Elbaum, 2004), ensuring traceability to the mole.

Studies of homogeneity, characterization, short-term stability, and long-term stability were performed. These studies and associated uncertainties were calculated according to procedures established by the ISO Guide 35:2006 (ISO Guide 35, 2006). The combined standard uncertainty (uCRM) of both CRM was calculated by combining the contributions of uncertainties of homogeneity, characterization and long-term stability studies, according to Equation 2.

$$u_{CRM} = \sqrt{u_c^2 + u_h^2 + u_s^2} \tag{2}$$

where:

u_c = characterization uncertainty;

u_h = homogeneity uncertainty;

u_s = long-term stability uncertainty.

The short-term stability uncertainty was not included in the combined standard uncertainty presented above, because it was studied only to define the transport conditions and it is negligible in relation to long-term stability uncertainty. The certified value presents the expanded uncertainty (U), and it is expressed as the product of the combined standard uncertainty (u_{CRM}) and the coverage factor k=2 (approximately 95% of confidence level) (Guimarães et al, 2014; Neves et al, 2015; Inmetro, 2012a; Inmetro, 2013b; Inmetro, 2015; Inmetro, 2016a).

Certification of BTEX CRM

According to the short-term stability study, the CRM of BTEX (CRM 8298.0001) must be transported for a maximum period of 50 days at 40 °C without changing its property values certified, and it should be stored in the dark at temperatures below 4 °C, but they cannot be frozen.

The certification was done by gas chromatography coupled mass spectrometry (ion trap) using the isotope dilution technique (GC/IT/IDMS).

The GC was equipped with a ZB-WAX (30 m x 0.32 mm x 0.5 µm) capillary column. Helium was used as a carrier gas at 1 mL/min constant flow rate. All injections were performed on split mode (1:50). The temperatures are 220 °C to injector, 220 °C to trap, 230 °C to transfer line and 50 °C to manifold. The oven temperature was programmed as follows: 40 °C for 1 min; 2 °C min^{-1} to 48 °C; 30 °C min^{-1} to 70 °C; 4 °C min^{-1} to 80 °C; 40 °C min^{-1} to 150 °C, with 10 min final hold. The operational ion-trap conditions: axial modulation voltage: 4.0 V; multiplier offset: 0 V; emission current: 10 µA; electron multiplier voltage: 1500 V. The injection volume was 1 µL.

The certified value and their expanded uncertainties for each compound present in the CRM 8298.0001 (BTEX) are presented in Table 3 (Inmetro, 2012a; Inmetro, 2016a; Neves et al, 2015).

Certification of PAH CRM

The CRM of PAH (CRM 8363.0001a) must be transported for a maximum period of 60 days at 30 °C without changing its property values certified, and it should be stored in the dark at temperatures below 4 °C, but they cannot be frozen.

The characterization was done by gas chromatography coupled mass spectrometry using the isotope dilution technique (GC-IDMS), using the exact matching for quantification. The GC was equipped with a VF-5ms (60 m x 250 mm x 0.25 µm) capillary column. Helium was used as a carrier gas at 1 mL min^{-1} constant flow rate. All injections were performed on splitless. The temperatures were 290 °C to

Table 3. Certified value and expanded uncertainty of CRM 8298.001 from Inmetro (Brazil)

VOC in Methanol	Certified Value (µg/g)	Certified Value (µg/mL)[a]
Benzene	34.1 ± 3.1	27.0 ± 2.5
Toluene	172.6 ± 4.5	136.7 ± 3.6
Ethylbenzene	87.2 ± 3.2	69.0 ± 2.5
ortho-xylene	87.3 ± 2.8	68.8 ± 2.2
meta-xylene	86.4 ± 2.4	68.4 ± 1.9
para-xylene	86.9 ± 4.1	69.1 ± 3.2

[a]The certified value in µg/mL units were obtained by multiplying the certified values in µg/g by the density of the solution of CRM of BTEX at 20 °C (0.791773 ± 0.000032 g/cm^3). The measurement density uncertainty was included in estimation of measurement uncertainty in µg/mL units.

injector, 250 °C to detector and 280 °C to interface. Mode Scan and SIM. The oven temperature was programmed as follows: 60 °C (2 min), 120 °C (20 °C/2 min), 180 °C (6,0 °C /2 min), 290 °C (3,0 °C/29,33 min). The injection volume was 1 µL. The certified value and their expanded uncertainties for each compound present in the CRM 8363.0001a (PAH) are shown in Table 4 (Inmetro, 2013b, Guimaraes et al, 2014).

After shelf-life of this CRM, a new certification study was realized and the batch received the code CRM 8363.0001b. Two compounds have only informative values, as presented at Table 5. The new certified values and their expanded uncertainties for 14 compounds present in the CRM 8363.0001b (PAH) is in Table 6 (Inmetro, 2015).

Bioethanol From Sugar Cane

Biofuels constitute a viable alternative in relation to petroleum-derived fuels and can be considered as a significant energy source of the future. It can decrease the consumption of petroleum derivate, contributing to the reduction of environmental impact, as well as minimize particulate material and gas emissions which causes the greenhouse effect. The participation of biofuels in the energetic matrix depends on a complex array of factors, involving many production aspects (raw material and technology) and the sustainability (social, economic and environmental) of this energy source (Fraga, Ribeiro, Sobral, Dias, Gonçalves, Borges & Gonzaga, 2012; Zuurbier & Vooren, 2008; Cortez, Stradiotto, Zanoni, Fraga & Borges, 2010).

Table 4. Certified value and expanded uncertainty of CRM 8363.0001a from Inmetro (Brazil)

PAH in toluene	Certified Value (µg/g)	Certified Value (µg/mL)[b]
Naphthalene	4.42 ± 0.18	3.83 ± 0.16
Acenaphthylene	4.91 ± 0.16	4.25 ± 0.13
Acenaphthene	4.86 ± 0.32	4.22 ± 0.28
Fluorene	4.94 ± 0.32	4.28 ± 0.27
Phenanthrene	4.97 ± 0.37	4.31 ± 0.32
Anthracene	5.12 ± 0.15	4.43 ± 0.13
Fluoranthene	5.04 ± 0.17	4.37 ± 0.15
Pyrene	5.10 ± 0.35	4.42 ± 0.30
Benz[*a*]anthracene	4.96 ± 0.22	4.30 ± 0.19
Chrysene	5.00 ± 0.36	4.33 ± 0.31
Benzo[*b*]fluoranthene	5.05 ± 0.34	4.37 ± 0.30
Benzo[*k*]fluoranthene	5.08 ± 0.30	4.41 ± 0.26
Benzo[*a*]pyrene	5.55 ± 0.73	4.81 ± 0.64
Indeno[*1,2,3-c,d*]pyrene	5.14 ± 0.73	4.46 ± 0.64
Dibenz[*a,h*]anthracene	4.96 ± 0.39	4.30 ± 0.34
Benzo[*ghi*]perylene	5.09 ± 0.35	4.41 ± 0.30

[b]The certified value in µg/mL units were obtained by multiplying the certified values in µg/g by the density of the solution of CRM of of PAH at 20 °C (0.8668562 ± 0.0000011 g cm^{-3}). The measurement density uncertainty was included in estimation of measurement uncertainty in µg/mL units.

Table 5. Informative value and expanded uncertainty of CRM 8363.0001b from Inmetro (Brazil)

PAH in toluene	Informative value (µg/g)	Informative value (µg/mL)[c]
Naphthalene	4.51 ± 0.15	3.91 ± 0.13
Dibenz[a,h]anthracene	4.50 ± 0.25	3.90 ± 0.22

[c]The informative value in µg/mL units were obtained by multiplying the informative value in µg/g by the density of the solution of CRM of PAH at 20 °C (0.8668562 ± 0.0000011 g cm-3). The measurement density uncertainty was included in estimation of measurement uncertainty in µg/mL units.

Table 6. Certified value and expanded uncertainty of CRM 8363.0001b from Inmetro (Brazil)

PAH in toluene	Certified value (μg/g)	Certified value (μg/mL)[d]
Acenaphthylene	4.86 ± 0.20	4.21 ± 0.17
Acenaphthene	4.77 ± 0.23	4.13 ± 0.20
Fluorene	4.85 ± 0.24	4.21 ± 0.21
Phenanthrene	4.96 ± 0.27	4.30 ± 0.24
Anthracene	5.01 ± 0.14	4.34 ± 0.12
Fluoranthene	5.04 ± 0.23	4.37 ± 0.20
Pyrene	5.08 ± 0.34	4.41 ± 0.29
Benz[*a*]anthracene	4.95 ± 0.19	4.29 ± 0.16
Chrysene	4.93 ± 0.31	4.27 ± 0.27
Benzo[*b*]fluoranthene	4.95 ± 0.33	4.29 ± 0.29
Benzo[*k*]fluoranthene	4.97 ± 0.27	4.31 ± 0.24
Benzo[*a*]pyrene	5.49 ± 0.72	4.76 ± 0.63
Indeno[*1,2,3-c,d*]pyrene	5.08 ± 0.73	4.41 ± 0.63
Benzo[*ghi*]perylene	5.07 ± 0.32	4.39 ± 0.28

[d]The certified value in μg/mL units were obtained by multiplying the certified values in μg/g by the density of the solution of CRM of PAH at 20 °C (0.8668562 ± 0.0000011 g cm^{-3}). The measurement density uncertainty was included in estimation of measurement uncertainty in μg/mL units.

In order to guarantee the bioethanol quality and its global marketability, it is necessary that the analytical measurements of the quality parameters of bioethanol (ANP, 2011) present metrological traceability and reliability which are obtained mainly by the use of certified reference material (CRM).

Since 2005, Inmetro has been studying the different parameters established in the specifications for bioethanol. It is presented that the certification of some parameters of the bioethanol batch are actually available.

Additionally, Inmetro participated in the European Project called BIOREMA (Biorema, 2010) in order to certify CRMs for bioethanol and biodiesel and coordinate an interlaboratory comparison in the world. In 2010, INMETRO has coordinated an interlaboratory comparison in anhydrous bioethanol for the Project "Standards for the Ethanol of Africa and Latin America" named PEAAL, in Portuguese. This Project was aimed to assist in technical qualification the National Metrology Institutes (NMI) and independent quality laboratories, indicated and invited by the NMIs, which are involved with the quality control of biofuels in their respective countries (PEAAL, 2010).

Certification Procedures

A batch of approximately 40 L of the raw material of bioethanol was produced with an ethanol content of 99.5% mass fraction, provided from a Brazilian producer. The bioethanol was bottled in approximately 200 amber glass bottles of 500 mL (for the determination of the total acid number, electrolytic conductivity and pHe parameters) and approximately 2000 amber glass ampoules of 10 mL (for the remaining parameters, as shown in Table 5). The studies for the certification of these parameters (the property values) for bioethanol samples were performed in accordance with ISO 17034 (ABNT NBR ISO 17034, 2017) and ISO Guide 35 (ABNT ISO Guia 35, 2012). Table 7 shows the methods used for certification of each parameter of bioethanol CRM.

The homogeneity study was performed for all parameters and all the measurements were done on the same day, with a storage temperature of 21 °C. The statistical test applied was the Analysis of Variance (ANOVA). The short-term stability study was carried out for five weeks, with a storage temperature of 50 °C and the long-term stability study, for 24 months, with a storage temperature of 21 °C. A shelf life of one year was proposed for this CRM.

Table 7. Equipment used for the analysis of each parameter of bioethanol CRM from Inmetro (Brazil)

Parameter	Equipment
Chloride	Ionic bidimensional chromatographic system with pre-concentration and detection system by conductivity suppression
Copper and Iron	Inductively Coupled Plasma Optical Emission Spectrometry(ICPOES) coupled to ultrasonic nebulizer and a membrane desolvator and the isotopic dilution using inductively coupled plasma-mass spectrometry (ICP-MS)
Sodium	Inductively Coupled Plasma Optical Emission Spectrometry (ICP OES) coupled to ultrasonic nebulizer and a membrane desolvator and atomic emission spectrometry with flame
Electrolytic conductivity	Conductivity meter coupled with one cell conductivity with cell constant of 0.090 cm^{-1}
Ethanol content	Gas chromatographic with flame ionization detector (GC FID) by on-column injection technique and propanol as internal standard
pHe	pH meter coupled with pH combined electrode with internal solution 3 M KCl
Sulphate	Ionic bidimensional chromatographic system with pre-concentration and detection system by conductivity suppression
Total acid number	Automatic potentiometric Titrator (Titrando) composed by automatic burettes Dosino 800
Water content	Karl Fischer coulometric titration with the Hydranal coulomat AG solution as reagent

The characterization studies approach is the most accurate method available for measurements in bioethanol since each parameter is certified in a prepared bath. All the measurements were done in six replicates for each parameter. The methods used in the characterization study presented metrological traceability with calibrated equipment and the use of a CRM. The uncertainty estimative of each parameter was calculated considering all the contributions that impacted the measurement methods.

Certification of Electrochemical Parameters of CRM of Bioethanol

According to the certificate of the last batch of the CRM of Ethanol fuel, developed by the Electrochemistry Laboratory from Dimqt/Inmetro, it should be stored at temperature of (20.0 ± 5.0) °C. It's recommended after use, to close the bottle and store it in the refrigerator, avoiding contact with possible contaminants (vapour acids, oxides and other gases). The minimum volume to be used is of 20 mL.

The determination of total acid number was performed by volumetric titration with potentiometric detection using an automatic potentiometric titrator. In order to obtain the titration curve and the end point, a combined glass electrode containing a 2.0 mol L^{-1} LiCl ethanolic filling solution was used. The potentiometric titrator coupled to the combined glass electrode was calibrated using CRMs of pH 6.9 (8855.0001) and pH 4.0 (8832.0001). The ethanol samples were titrated with a NaOH 0.02 mol L^{-1} solution that was standardized with a CRM of HCl 0.01 mol kg^{-1} (08.2/10.0003, Inmetro) certified by the primary coulometry system from Inmetro. The result is reported as acetic acid mass, in mg, per ethanol volume, in L.

The primary system of electrolytic conductivity from Inmetro was used for the characterization of the property value (Jensen, 2009). The electrolytic conductivity was measured using a conductivity meter coupled to a conductivity cell with a cell constant of about 0.1 cm^{-1} for the homogeneity and stability studies. A CRM of 5 μS cm^{-1} (8435.0002, Inmetro) was used for the calibration of the conductivity meter (Inmetro, 2013c). All the measurements were carried out at 25 °C.

For the homogeneity study the pH was carried out using a pH meter coupled to a combined glass electrode containing a 2.0 mol L^{-1} LiCl ethanolic filling solution. The system was calibrated using CRM of pH 6.9 (8855.0001) and pH 4.0 (8832.0001). For characterization study the pH was performed using two different pH meter coupled to the respective combined glass electrodes containing the same filling solution described above. All the measurements were carried out at 25 °C. The certified value and their expanded uncertainties for each compound present in the CRM 8313.0004 (Ethanol fuel) are presented in Table 8 (Inmetro, 2013d).

Table 8. Certified value and expanded uncertainty of CRM 8313.004 from Inmetro (Brazil) –electrochemical parameters

Parameter	Certified Value and Expanded Uncertainty	Unit
Total acidity	7.6 ± 1.1	mg L^{-1}
Electrolytic conductivity	1.20 ± 0.10	μS cm^{-1}
pH	6.51 ± 0.34	-

Certification of Organic Parameters of CRM of Bioethanol

According to the certificate of the last batch of the CRM of bioethanol fuel, the minimum sample size of the CRM to be used for determination of water content is 0.025 g and for determination of ethanol content is 0.16 g. This CRM should be stored at temperature of (22.0 ± 2.0) °C and after its use should be discarded.

The water content was determined by means of Karl Fischer coulometric titration with the Hydranal coulomat AG solution as reagent; a current generator electrode of 400 mA and a platinum indicator electrode. The tension range applied was from 50 to 70 mV. The titration was started after the stabilization of the equipment with a drift of 20 μgmin^{-1} on 120 s. The titration curve was done with the measurement of the points each 2 s of the analysis. Each sample was transferred into the sealed flask, the aliquots were withdrawn, weighted and injected in the equipment. The determinations were done in triplicate. The same procedure was applied in a previous lot (Rodrigues, Guimarães, Sousa, Silva, Fraga, Souza & Cunha, 2012).

The determination of the ethanol content was carried out by the use of the gas chromatograph with flame ionization detector (GC-FID) by on-column injection technique and propanol as internal standard. Samples of bioethanol were weighted in glass recipient of 5 mL and transferred to a flask contained 250 g of water. The solutions were homogenized during 12 h. One aliquot of this solution was diluted 1:1 with internal standard solution in a glass flask. The ethanol content was determined by calibration curve equation that was obtained by square minimum method, OLS - Ordinary Linear Square. The determinations were done in duplicate.

The certified value and their expanded uncertainties for each compound present in the CRM 8314.0002a (Bioethanol fuel) are presented in Table 9 (Inmetro, 2013d).

Table 9. Certified value and expanded uncertainty of CRM 8314.002a from Inmetro (Brazil) – organic parameters

Parameter	Certified Value and Expanded Uncertainty	Unit
Water content	0.257 ± 0.019	w/w (mass fraction)
Ethanol content	99.2 ± 2.0	w/w (mass fraction)

Certification of Inorganic Parameters of CRM of Bioethanol

The Inorganic Laboratory from Dimqt/Inmetro developed 02 (two) certificates for different inorganic parameters of the CRM of Bioethanol.

The copper and iron content were determined using an ICP OES coupled to ultrasonic nebulizer and membrane desolvator. The water content of the calibration solutions was between 0.25-0.40% (m/m). These methodologies were validated comparing the results with the isotopic dilution using inductively coupled plasma-massa spectrometry (ICP-MS). The results of measurements of copper and iron content parameter have metrological traceability to SRM 3114 and SRM 3126a from NIST, and IRMM-14 (Fe (56)) and IRMM-632 (Cu (65)).

The content of sodium was certified combining the results obtained from ICP OES coupled to ultrasonic nebulizer and membrane desolvator and atomic emission spectrometry with flame. The results of measurements of sodium content parameter have metrological traceability to SRM 3152 from NIST.

Both certificates of those parameters of this CRM are not available on Inmetro's website anymore as their validity time is already expired. Nevertheless, their certified values and respective expanded uncertainties are presented in Table 10 (Inmetro, 2012b; Inmetro, 2010).

Table 10. Certified value and expanded uncertainty of CRM 8315.001 from Inmetro (Brazil) – Inorganic parameters

Parameter	*Certified Value and Expanded Uncertainty* (mg/kg)
Copper	0.149 ± 0.007
Iron	0.0186 ± 0.0014
Sodium	0.83 ± 0.06
Sulphate	1.152 ± 0.096

CONCLUSION

Considering the importance on reliability and quality of measurements, the certified reference materials related to environmental area developed by the National Metrology Institute from Brazil, Inmetro are presented. So, this work presents as novelty and interest to readership, the use of primary methods and achievements made in developing of environmental CRM, in order to support the needs of the national industry and to disseminate traceability among the society. We believe this paper will make the scientific community aware of the importance of reliability and traceability in environmental measurements by the use of this kind of certified reference materials.

Nowadays, in cities, vehicle emissions play an important role in the level of air pollution. So, the analysis of air quality using this type of CRM by responsible agencies requires reliable and comparable analytical methods in order to reduce the emission of such gases. Besides, when comparing, for instance, a national CRM of a gas mixture, such as automotive emission, and an international mixture developed at the same conditions, the price for the national CRM is two times cheaper than from a foreigner NMI.

Among the organic compounds, the volatile organic compounds are highlighted – BTEX, and the Polycyclic Aromatic Hydrocarbons (PAH), which are environmental pollutants found in water, air and soil and they are considered of interest to public health due to environmental persistence and their potential toxicity (mutagenic and carcinogenic). For this reason, the analysis of these environmental contaminants with a metrological focus is considered essential, and developing their CRM is considered of essential significance.

The bioethanol also plays an important role as source of alternative energy. It is a renewable and less pollutant source, as it reduces the environmental impact caused by greenhouse gases and particulate material emissions.

REFERENCES

ABNT. (2012). [– *Materiais de Referência – Princípios gerais e estatísticos para certificação.* Rio de Janeiro: ABNT.]. *ABNT ISO Guia, 35*, 2012.

ABNT. (2016). [– *Materiais de Referência – Termos e Definições selecionados.* Rio de Janeiro: ABNT.]. *ABNT ISO Guia, 30*, 2016.

ABNT. (2017). [– *Requisitos gerais para a competência de produtores de materiais de referência.* Rio de Janeiro: ABNT.]. *ABNT NBR ISO, 17034*, 2017.

Agência Nacional de Petróleo (ANP). (2011). *Resolução ANP N° 7, Brasília.* Retrieved July, 2013, from http://nxt.anp.gov.br/NXT/gateway.dll/leg/resolucoes_anp/2011/fevereiro/ranp7-2011.xml

Cortez, L. A. B., Stradiotto, N. R., Zanoni, M. A. B., Fraga, I. C. S., & Borges, P. P. (2010). *Sugar cane bioethanol R&D for productivity and sustainability, 4th Part, 21.* São Paulo: Ed.Blücher.

Duewer, D. L., Parris, R. M., White, E., May, W. E., & Elbaum, H. (2004). An Approach to the Metrologically Sound Traceable Assessment of the Chemical Purity of Organic Reference Materials. NIST Special Publication 1012.

EURACHEM / Cooperation on International Traceability in Analytical Chemistry (CITAC). (2000). Guide CG 4. In Quantifying uncertainty in analytical measurement (2nd ed.). London: Author.

Executive Summary from BIOREMA Reference Materials for Biofuel specifications. (2010). Retrieved July, 2013, from http://www.inmetro.gov.br/painelsetorial/palestras/publishable-summary.pdf

Fraga, I. C. S. F., Ribeiro, C. M., Sobral, S. P., Dias, J. C., Gonçalves, M. A., Borges, P. P., & Gonzaga, F. B. (2012). Certified reference material of bioethanol for metrological traceability in electrochemical parameters analyses. *Talanta, 99*, 99–103. doi:10.1016/j.talanta.2012.05.024 PMID:22967527

Guimarães, E. F., Rego, E. C. P., Cunha, H. C. M., Rodrigues, J. M., & José Daniel Figueroa-Villar, J. D. (2014). Certified Reference Material for Traceability in Environmental Analysis: PAHs in Toluene. *Journal of the Brazilian Chemical Society, 25*(2), 351–360.

Guimarães, E. F., Vieira, A. A., Rego, E. C. P., Garrido, B. C., Rodrigues, J. M., & Figueroa-Villar, J. D. (2015). Quantitative nuclear magnetic resonance for purity assessment of polycyclic aromatic hydrocarbons. *Metrologia, 52*(4), L15–L22. doi:10.1088/0026-1394/52/4/L15

Inmetro. (2010). *Dimci 2354/2010, Certificate of CRM 8315.0001: Ethanol Fuel, Brazil*. Author.

Inmetro. (2012a). *Dimci 1477/2010c, Certificate of CRM 8298.0001: BTEX in methanol, Brazil*. Author.

Inmetro. (2012b). *Dimci 0009/2012, Certificate of CRM 8313.0002: Bioethanol Fuel, Brazil*. Author.

Inmetro. (2013a). *Dimci 1538/2013, Certificate of CRM 8634.0035: Automotive emissions in nitrogen, Brazil*. Author.

Inmetro. (2013b). *Dimci 0142/2013, Certificate of CRM 8363.0001a: PAH in toluene, Brazil*. Author.

Inmetro. (2013c). *Dimci 1384/2013, Certificate of CRM 8435.0002: Electrolytic conductivity of 5 $\mu S\ cm^{-1}$, Brazil*. Author.

Inmetro. (2013d). *Dimci0792/2013, Certificate of CRM 8313.0004: Ethanol Fuel, Brazil*. Author.

Inmetro. (2013e). *Dimci 2526/2012, Certificate of CRM 8314.0002a: Ethanol Fuel, Brazil*. Author.

Inmetro. (2015). *Dimci 02715/2014a, Certificate of CRM 8363.0001b: PAH in toluene, Brazil*. Author.

Inmetro. (2016a). *Dimci 1477/2010g, Certificate of CRM 8298.0001: BTEX in methanol, Brazil*. Author.

Inmetro. (2016b). *Dimci 2526/2012c, Certificate of CRM 8314.0002a: Ethanol Fuel, Brazil*. Author.

Interlaboratory Comparison in Anhydrous Bioethanol. (n.d.). Retrieved July, 2013, from http://www.inmetro.gov.br/metcientifica/pdf/Protocol-Bioethanol-PEAAL.pdf

International Agency for Research on Cancer (IARC). (2010). *Complete list of agents, mixtures and Exposures Evaluated and Their Classification*. Retrieved July, 2013, from http://www.iarc.fr

International Organization for Standardization. (2001). ISO 6143 - Gas analysis Comparison methods for determining and checking the composition of calibration gas mixtures. Geneva: ISO.

International Organization for Standardization. (2015). ISO 6142 - Gas analysis Preparation of calibration gas mixtures Gravimetric methods. Geneva: ISO.

Jensen, H. D. (2009). *Final Report on CCQM-K36.1*. Metrologia, 46, Tech. Suppl., 08004.

Joint Committee for Guides in Metrology (JCGM). (2008). *Evaluation of measurement data Guide to the expression of uncertainty in measurement*. Retrieved July, 2013, from http://www.bipm.org/utils/common/documents/jcgm/JCGM_100_2008_E.pdf

Milton, M. J. T., & Quinn, T. J. (2001). Primary methods for the measurement of amount of substance. *Metrologia*, *38*(4), 289–296. doi:10.1088/0026-1394/38/4/1

Neves, L. A., Almeida, R. R., Rego, E. P., Rodrigues, J. M., Carvalho, L. J., & Goulart, A. L. M. (2015). Certified reference material of volatile organic compounds for environmental analysis: BTEX in methanol. *Analytical and Bioanalytical Chemistry*, *407*, 3225–3229. doi:10.100700216-014-8314-6 PMID:25450051

Rodrigues, J. M., Guimarães, E. F., Sousa, M. V. B., Silva, V. F., Fraga, I. C. S., Souza, V., & Cunha, V. S. (2012). Certified Reference Material to water content determination in bioethanol fuel. *Quimica Nova*, *35*(5), 1011–1015. doi:10.1590/S0100-40422012000500026

van der Veen, A.M.H., Oudwater, R., Ziel, P.R., van Wijk, J.I.T., Augusto, C.R., Fioravante, A.L., … Durisova, Z. (2011). EURAMET QM-S4 Final Report – Key comparison automotive gas mixture. *Metrologia*, *48*.

Zuurbier, P., & Vooren, J. V. (2008). *Contributions to climate change mitigation and the environment*. Wageningen Academic Publishers.

Chapter 8
The Challenge of Developing Primary Standards Mixtures of Carbon Dioxide at Atmospheric Levels:
Establishing Traceability to Ghg Monitoring Analysis

Andreia de Lima Fioravante
National Institute of Metrology Standardization and Industrial Quality, Brazil

Cristiane Rodrigues Augusto
National Institute of Metrology Standardization and Industrial Quality, Brazil

Claudia Cipriano Ribeiro
National Institute of Metrology Standardization and Industrial Quality, Brazil

Valnei Smarçaro da Cunha
National Institute of Metrology Standardization and Industrial Quality, Brazil

Luiz Antônio d'Avila
Universidade Federal do Rio de Janeiro, Brazil

Thays Silva Ximenes
National Institute of Metrology Standardization and Industrial Quality, Brazil

DOI: 10.4018/978-1-5225-5406-6.ch008

ABSTRACT

In recent years, the need for the quality control, reliability, and traceability of analytical results has been strongly emphasized. The National Metrology Institute – Inmetro is involved in improving measurement capabilities to provide confidence level regarding greenhouse gases standards. The chapter describes the development of certified reference materials of carbon dioxide in atmospheric synthetic air that are used to monitor its concentration in the atmosphere. Considering the contributions from gravimetry preparation and its verification analysis by cavity ringdown spectroscopy (CRDS), the relative expanded uncertainty of the standard mixture of carbon dioxide was lower than 0.5% for the range studied from 370 to 550 µmol/mol, which is comparable to the average of 0.25% relative uncertainties presented on international standards mixtures.

INTRODUCTION

Greenhouse gases, as defined by ISO 14065 Clause 3.1.1, are gaseous constituent of the atmosphere, both natural and anthropogenic, that absorbs and emits radiation at specific wavelengths within the spectrum of infrared radiation emitted by Earth's surface, the atmosphere, and clouds. Include carbon dioxide (CO_2), methane (CH_4), nitrous oxide (N_2O), hydrofluorocarbons (HFCs), perfluorocarbons (PFCs) and sulfur hexafluoride (SF6) (ISO14065, 2013).

It is well known that improved quality and monitoring points in greenhouse gases (GHG) is needed to better predict climate tendencies and monitor sources and sinks of GHG, but also there is the need to increase certainty in achieved values of the mitigation reduction goals of GHG. All of these are presented in the Intended Nationally Determined Contributions (INDC), which is a term used under the United Nations Framework Convention on Climate Change (UNFCCC) for reductions in GHG that all countries that signed the UNFCCC were asked to publish in the lead up to held in Paris (United Nations, FCCC, 2015).

Implementing efficient provisions in order to protect biodiversity and to act against climate change requires an effective and efficient, demand-oriented quality infrastructure (QI). In order to support the transition process towards a "green economy", there is a need to focus on three key areas regarding environment: study of ISO standards and development of standards, improvement of the metrological system, and validation programs, verification and certification. Regional research laboratories worldwide will benefit from reliable, accredited measurement and testing capacities which have to be provided to ensure sustainable management.

Then, building reliable testing of GHG based on new capabilities of traceable measurement results is needed, both for atmospheric measurements and local measurement of emission factors of GHG. Besides, environmental legislation requires international standards of comparison.

Inmetro, is a locus of knowledge and credibility who plays an indispensable role, mainly towards the development and maintenance of certified reference material (CRM), as well as for carrying and spread the SI units of measurement, and its harmonization on a global level, providing traceability to a large number of users via a metrological chain. The Gas Analysis Laboratory (Lanag) from the Chemistry and Thermal Metrology Division (Dimqt) from Inmetro has the knowledge and facilities on producing certified gas mixtures, both binary and multicomponent CRM [van der Veen et *al.*, 2011]. However, for atmospheric levels of greenhouse gases, especially carbon dioxide and methane, Inmetro is still developing such standards, and since there is a demand for this CRM, this study is very promising and will also help to reduce the importation costs.

Within this framework the assurance of the required traceability of all measurements reflects in the use of reference materials which ensures that the results obtained by different laboratories are comparable and traceable. This reliability is essential in the maintenance of a universal and consistent system of measurements among different laboratories, besides disseminating traceability throughout society (Fioravante et al., 2013).

PRIMARY STANDARD GAS MIXTURES OF GHG REFERENCE MATERIAL

The use of reference mixtures/materials with matrix compositions and analytes concentrations similar to those found in environmental samples makes it possible to obtain reliable measurement results. Investigations of samples of different types of gas mixtures are being carried out on an ever greater scale, so there is an increasing need for new techniques in preparing primary standard gas mixtures (PSM).

Primary Reference Materials (PRM) are prepared gravimetrically as Primary Standard Gas mixtures (PSM) by National Metrology Institutes (NMI) using guideline ISO 6142:2015 - "Gas analysis – Preparation of calibration gas mixtures – Weighing methods". The accuracy level of PRM is the highest commercially available mixture, and therefore, not mended for direct use as working standard calibrating gas analysers, but more as reference standard to which working standards are traceable to.

Certified Reference Materials (CRM) are prepared in compliance with ISO 17034:2016 – "General requirements for the competence of reference materials producers", and are certified by the national standards laboratory using guideline ISO 6143:2001 - "Gas analysis – Determination of composition of calibrating gas mixtures – Comparison methods". In this comparison the method of multipoint calibration is applied, where the analyser is calibrated using PRM.

The accuracy level of CRM is less than a PRM, and therefore, CRM are mend for daily calibration of analysers or as reference point for other mixtures (traceability). In addition, there are reference materials (RM), which are gas mixtures with sufficiently well known properties in order to use them to verify an instrument or a measurement method (RvA-I6.2, 2004).

Standard gas mixtures are used mainly in validating analytical procedures and for assessing the competence of laboratory staff. They are also essential in monitoring air pollution. The increasing need to monitor the degree of air pollution has driven the search for new techniques of preparing PSM containing reliable mixtures with analytes at ever smaller concentrations levels [e.g., ppb and ppt], closely resembling those of environmental samples. However, it challenges analysts to produce such reference gaseous mixtures (Fioravante et al., 2017).

Carbon Dioxide in Atmospheric Synthetic Air Mixtures

Advance in specific metrological services of providing local CRM mixtures to account GHG based on metrologically traceable measurement results are increasingly being demanded, especially for the three main greenhouse gas contributors: carbon dioxide, methane and nitrous oxide. Improve robustness and accuracy of climate data and uncertainties associated with measurements of GHG in outdoor measurements and in local emission factors of specific GHG are needed to have more certainty in the achieved values of INDC. Detected local needs of user-regulator relationships will be benefit from the use of CRM of GHG, therefore it will be possible the measurement of local GHG emission factors to support with metrology the achievement of the INDCs in the respective countries. It's essential to have local traceability to the GHG scale established by the National Oceanic and Atmospheric Administration (NOAA), who acts as the Central Calibration Laboratory (CCL) for such gases in the GAW-WMO program; this study allows to supply local needs of atmospheric stations that can take part of the GAW-WMO program of monitoring of GHG or any other research activity on this subject.

Thus, making available local CRM and their dissemination to end users in Latin America countries for new services of measurement of GHG based on traceable measurements should be encouraged. The following services are provided by NMI, such as calibration of gas mixtures reference materials (CGM) from customers, like the gas industries, or the certification of gas mixtures, in order to supply reference materials for customers. These services can be judged as new and innovative for mixtures due to the nature of the end applications to be promoted. That is, specific from sector/application/country/local emission factors with lower uncertainties to those of the IPPC default values to support the certainty of INDC reduction targets, and availability of local CGM compatible with the NOAA-GHG scale as calibration centre of the GAW-WMO needed for atmospheric monitoring community.

This new candidate for certified reference material (CRM) being developed by the Laboratory of Gas Analysis (Lanag) from the Chemistry and Thermal Metrology Division (Dimqt) of Inmetro consists in a primary reference mixture (PRM) of carbon dioxide (CO_2) in a matrix of synthetic air at atmospheric levels. The standard mixtures developed by renomated NMI, such as NIST and KRISS, use as matrix whole dry air sampled from clean dry air at specific places. This paper presents the development of a standard using a synthetic matrix of atmospheric air at the same level concentration as the whole air, being defined here as atmospheric synthetic air mixtures or by synthetic clean dry air (SCDA).

Gravimetric Development Procedure

The gravimetric preparation, as well as the calculation of the gas composition and associated uncertainty evaluation should be done in accordance with ISO 6142:2015. Before the gravimetric preparation of the mixture, the cylinder is cleaned by evacuation in a Pfeiffer turbo molecular pumping station followed by filling, after a sequence of pressure and vacuum, with high purity nitrogen (N_2) 99.9999% mol/mol. Aluminium gas cylinders of 5L treated with a proprietary process Aculife IV, by Scott Specialty Gases, were used. Afterwards, the empty cylinder is weighed in a mass comparator, Mettler balance, where it is compared against an identical (reference) cylinder. Then the pure gas or a pre-mixture with higher concentration of carbon dioxide in air is transferred via a filling station to the empty cylinder. This process of weighing and filling is repeated for all components of the matrix at the atmospheric air composition, i.e.: 0.9% mol/mol of argon, 21% mol/mol of oxygen and 79% mol/mol of nitrogen. Once the mixture is produced the final composition is calculated according to ISO 6142 by using the added masses of each component that is expressed as mole fraction µmol/mol.

The carbon dioxide in atmospheric synthetic air mixtures (CO_2/SCDA) were prepared in the range of 370 to 550 µmol/mol from the sequence dilution of a 3.0

(pre-mixture PM1) and a 0.5 (pre-mixture PM2) cmol/mol CO_2/SCDA pre-mixtures, preliminary prepared from the following pure gases, all of them from the supplier White Martins: CO_2 99.9995%, Argon 99.9999%, Oxygen 99.9999%, and Nitrogen 99.9999%. The scheme of these gravimetric planning preparation can be observed at Figure 1.

Primary gravimetric gas standards (PSM) are prepared in-house using parent gases with certified purity and contain no species that would affect the certified concentration values or be detrimental to the stability of the mixtures. The purity of the parent gases are checked against the manufacturers specification using high-sensitivity instrumentation, to ensure that no chemical or physical reactions take place during any stages of the preparation and dilution process, and to demonstrate the long term stability of the standards after preparation. Two main types of instrument are generally used for the analysis of such level of concentration of those standards studied:

- A number of gas chromatographic instruments with different detectors to enable quantify a wide range of gases to be measured with high sensitivity,

Figure 1. Layout/scheme of dilution gravimetric preparation of carbon dioxide in atmospheric synthetic air PSM

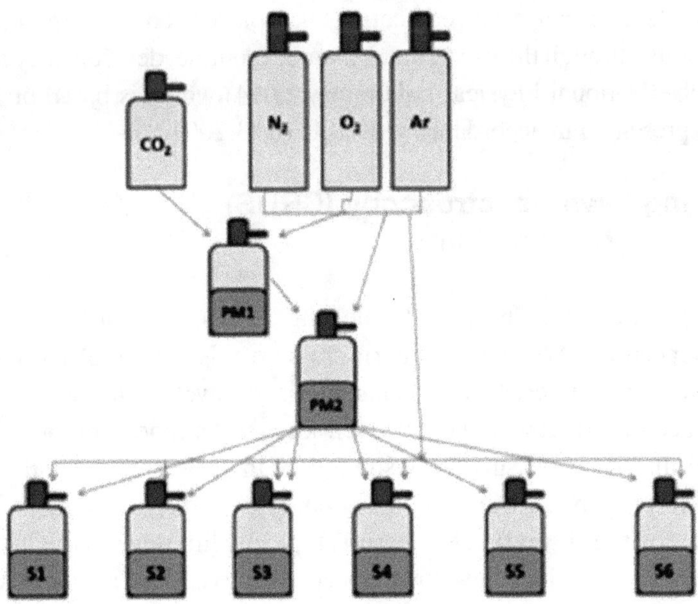

such as pulsed discharge of helion detector (PDHID), and flame ionization detector (FID) coupled to a methanizer catalyst; and

- Cavity ringdown spectroscopy system (CRDS).

For this work, first it was just applied the purity estimation based on ISO 6142, which uses the supplier data from gas industry of the pure gases to estimate the uncertainty of each component detected as impurity that can be interfering at the main component, i.e. trace levels of impurity carbon dioxide in pure nitrogen, for example. Afterwards, a purity analysis using GC/PDHID is intended to be done in order to confirm the amount of impurities in the pure gases used by the use of new standards of trace level compounds prepared for this specific analysis, such as trace levels of carbon monoxide and argon in pure nitrogen, nitrogen and oxygen in pure oxygen, and nitrogen and oxygen in pure argon.

The primary gas standards prepared are checked soon after the gravimetric preparation to confirm that their concentrations are as expected. After the gravimetric concentration of the mixtures produced were determined, the value was validated by comparing it with other primary standard (300 and 390 μmol/mol), developed by the Holland NMI. The analytical method employed in this paper was cavity ringdown (CRDS) by using a Tiger Optics – Prismatic equipment. The analysis were performed in 03 (three) different days under repeatability conditions with 10 (ten) runs of half an hour for stabilization each day. Based on the calibration curve, analysis values and its respective uncertainties were assigned to the amount of molar fractions of carbon dioxide. The analytical verification takes place to confirm gravimetric molar fraction results, through the use of the software Xgenline, developed by the England NMI – NPL, (National Physical Laboratory, 2014) which is based on calculation regression presented in accordance with ISO 6143:2001.

Cavity Ringdown Spectroscopy (CRDS) and Its Method Validation

Nearly every small gas-phase molecule (e.g., CO_2) has a unique near-infrared absorption spectrum. At sub-atmospheric pressure, this consists of a series of narrow, well-resolved, sharp lines, each at a characteristic wavelength. Because these lines are well-spaced and their wavelength is well-known, the concentration of any species can be determined by measuring the strength of this absorption, i.e. the height of a specific absorption peak. But, in conventional infrared spectrometers, trace gases provide far too little absorption to measure, typically limiting sensitivity to the parts per million at best. Cavity Ring-Down Spectroscopy (CRDS) avoids this sensitivity limitation by using an effective pathlength of many kilometers. It enables gases to be monitored in seconds or less at the parts per billion level.

Thus, CRDS works by attuning light rays to the unique molecular fingerprint of the sample species. By measuring the time it takes the light to fade or "ring-down", you receive an accurate molecular count in milliseconds. The time of light decay, in essence, provides an exact, non-invasive, and rapid means to detect contaminants in the air.

Regarding the principle of the CRDS to generate the result expected, a computer-controlled system tunes the laser off the absorption peak for the sample species to determine the τ empty value, equivalent to a zero-baseline correction. It tunes back to the absorption peak to determine the τ value, dependent on the sample species concentration. Based on Beer's Law, this value constitutes an absolute measurement and is unaffected by losses outside the ring-down cavity. The graph presented at Figure 2 depicts the concept of ring-down decay within the cavity after the laser source is shuttered. As the laser light bounces back and forth between the ultra-high reflective mirrors, the sample species absorbs the light energy until it's all gone. The CRDS schematic is presented at Figure 3.

Science should be used to improve quality of life. In this regard, scientists must guarantee the quality of their results. These results must provide relevant information, not just for the scientific community, but also for citizens. Demonstration of the ability of an analytical method to provide reliable results is of great importance to ensure quality, safety and efficacy in the analysis. Consequently, before an analytical method is implemented for routine use, it must be previously validated to demonstrate that it is suitable for its intended purpose. The final goal of the validation of an

Figure 2. Ring-down graph

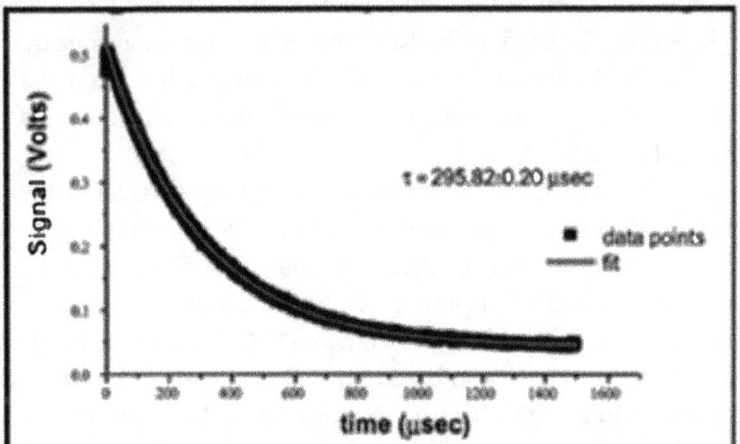

Figure 3. CRDS scheme

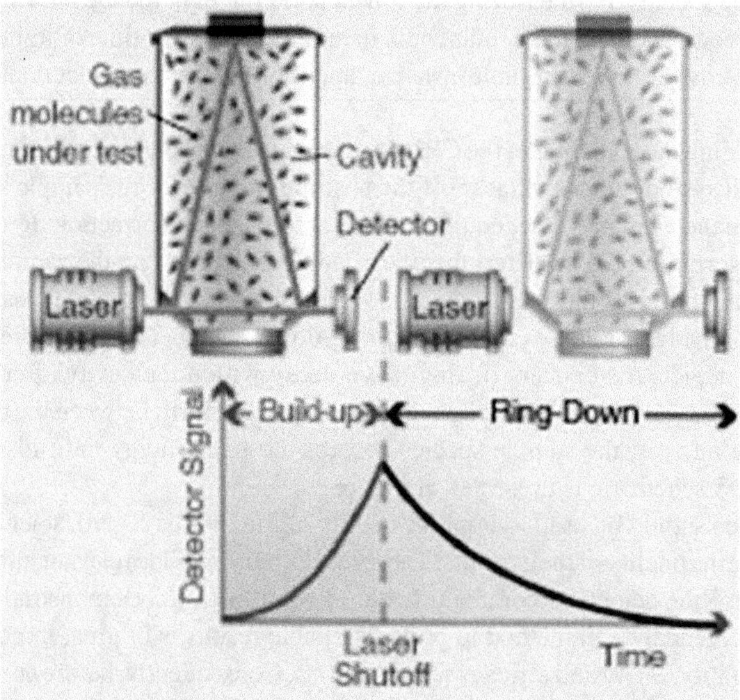

analytical method is to ensure that every future measurement in routine analysis will give a true assessment of the actual value of the content of an analyte in the sample. It is essential that laboratories have objective means and criteria to validation, that the test methods they perform lead to reliable results towards the desired quality. When employed standardized methods, it´s necessary to demonstrate that it has the appropriate manner, within the specific installation and facilities from the laboratory before deploying them, not needing to perform all the parameters from validation (Fioravante et al., 2018).

In the case of this work, the parameters designated to characterize the validation of each method developed was: selectivity, linearity, working and linear range, acceptability assessment, such as detection and quantification limits, trend and regression analysis, and comparison of the precision between the methods, according to what established by the document DOC-CGCRE-08 (Inmetro, 2010). Calculations of precision are also based on ISO standards 5725:1994 part 2, and uncertainty estimation derived from regression of the calibration curve adopted was based on ISO 6143:2001 using the software XGenline from NPL (National Physical Laboratory, 2014). The limit of detection and quantification were calculated according to the International Conference on Harmonization of Technical Requirements for Registration of Pharmaceuticals for Human Use (ICH).

Characterization of the Primary Standard Mixture

The final uncertainties obtained by the combination of just the gravimetry and verification analysis were compared to a certified uncertainty from a primary reference mixture (PRM) of the Dutch NMI, which is available on market.

Finally, after verification analysis, standard mixtures are then allowed to stand for a defined period before final certification is carried out. Following this period, all standards will be then certified.

This work presents the steps of developing a PSM of CO_2/SCDA with its results for gravimetry and verification analysis, and their respective uncertainties contributions. Besides, the results obtained for the method validation of the analysis on CRDS made for those type of standards are also presented.

These uncertainty results from the CRDS method validation can be compared to the preliminary results from Inmetro derived from an internation comparison for analysis of carbon dioxide at synthetic air at 380 μmol/mol analysed by gas chromatography.

RESULTS AND DISCUSSION

The gravimetric preparation data with the respective uncertainty was obtained by the Gas Metrology software developed by the Dutch NMI – VSL(VSL, 2006); this software was developed under demand to be used together with the gravimetric facilities based on calculation required by ISO 6142 and validated by all KC that the Gas Analysis Laboratory Lanag from Inmetro has participated before under the auspices of BIPM/CCQM. Results for the preliminary impurity analysis based on statistic estimation of the pure gas supplier quality information are presented below at Table 1. The gravimetric concentration and its uncertainty is given at Table 2.

With the method developed for the CRDS technique, calibration curves were drawn for each selected range, using the primary standards produced in-house to compose the curve and as samples, as well as, certified reference materials to verify the adjust of the proposed calibration curve. In this way, the establishment of the calibration curve that represents the set of points (xi, yi), which the known concentrations used are xi, are plotted against the responses of the instrument, yi,

Table 1. Purity analysis of pure gases used in the PSM batch preparation

	Carbon Dioxide 99.9995%		Argon 99.9999%		Oxygen 99.9999%		Nitrogen 99.9999%	
Component	**Concentration (mol/mol)**	**Relative Uncertainty (%)**	**Concentration (mol/mol)**	**Relative Uncertainty (%)**	**Concentration (mol/mol)**	**Relative Uncertainty (%)**	**Concentration (mol/mol)**	**Relative Uncertainty (%)**
N_2	---	---	$0.05x10^{-6}$	58	$0.275 x10^{-6}$	58	0.9999994	$2x10^{-5}$
Ar	---	---	0.99999885	0.00000041	$0.15 x10^{-6}$	58	---	---
CO	---	---	---	---	$0.05x10^{-6}$	58	$0.05x10^{-6}$	58
CO_2	0.9999984	$6.5x10^{-5}$	---	---	$0.05x10^{-6}$	58	---	---
O_2	$1.0x10^{-6}$	$5.8x10^{-5}$	$0.10 x10^{-6}$	58	0.999999125	$2x10^{-5}$	$0.25 x10^{-6}$	56
H_2O	$0.5 x10^{-6}$	58	$0.05x10^{-6}$	58	$0.25 x10^{-6}$	56	$0.25 x10^{-6}$	56
THC	$0.01x10^{-6}$	60	$0.05x10^{-6}$	58	$0.05x10^{-6}$	58	$0.05x10^{-6}$	58

Table 2. Gravimetric results for the curve batch of PSM CO_2/CDA

Mixture Code	Primary Standard Mole Fraction (μmol mol^{-1})	Gravimetric Uncertainty (%)
S1	370.55	0.02
S2	390.60	0.02
S3	421.32	0.02
S4	453.33	0.02
S5	479.94	0.02
S6	545.71	0.02

[1]Unidade em cmol/mol (%mol/mol).

Table 3. Gravimetric results and its purity table for the produced sample of PSM CO_2/CDA

Component	Primary Standard Mole Fraction (μmol mol^{-1})	Gravimetric Uncertainty (μmol mol^{-1})
CO_2	***380.03***	***0.08***
N_2	78.258[1]	0.001[1]
O_2	20.9014[1]	0.0008[1]
Ar	0.8025[1]	0.0005[1]
CO	0.05	0.02
THC	0.05	0.02
H_2O	0.25	0.10

[1]Unidade em cmol/mol (%mol/mol).

obtained at independent and repetitive conditions. A linear relationship between the concentrations and the measurement results is adjusted, obtained by the mathematical model of correlation: least squares numerical method.

With gravimetric concentrations and uncertainties, 05 (five) standards were selected to compose a linear model curve approaching the whole range studied for verification analysis according to ISO 6143. The analytical results obtained by cavity ringdown spectroscopy system (CRDS) were applied to get an adjusted fit model where the analytical or verification concentration and its respective uncertainties were obtained through a statistical regression. One PSM was randomly selected as the sample to be tested in the stipulated curve. Appling ISO 6143 calculation for concentration and uncertainties derived by this calibration curve analysis, results of the verification analysis of the primary standards prepared can be observed at Table 4 by means of deviation of gravimetric concentration and analysis value, as well as the relative combined expanded uncertainty for gravimetric and analysis verification.

According to the verification requirement stated by ISO 6142, all gravimetric mixtures results are considered approved when the absolute difference between the gravimetry and verification analysis results are lower than expanded combined uncertainties, as presented at equation 1.

$$\left|x_g - x_v\right| \leq k \times \sqrt{x_g^{\,2} + x_v^{\,2}} \tag{1}$$

where,

x_g – gravimetric concentration (μmol/mol)
x_v – verification analysis concentration (μmol/mol)
u_g – gravimetric absolute uncertainty (μmol/mol)
u_v – verification analysis absolute uncertainty (μmol/mol)

The relative uncertainty estimated by the calibration for the selected sample at the inferior limit of the range, 380 μmol/mol, was lower than 0.5%, which is accepted considering that the purity analysis was done only by estimation on supplier data quality. Therefore, all standards of CO_2/SCDA developed were considered approved during this step of verification analysis.

The validation of the CRDS analysis was performed using independent results under conditions of repeatability (5 replicates) and reproducibility (3 different

Table 4. Verification results for the curve batch of PSM CO_2/CDA

Mixture Code	Δ (%)[1]	U (%)[2]
S1	0.01	0.18
S2	0.17	0.48
S3	0.21	0.21
S4	0.03	0.09
S5	0.02	0.11
S6	0.01	0.14

[1]Relative deviation between the gravimetric and verification analysis concentration
[2]Relative combined expanded uncertainty (gravimetry and verification analysis)

days). It was selected 03 (three) primary standard gaseous mixtures produced by Lanag, which range of concentration adopted was from 370 to 420 μmol mol^{-1}, due to the specifications from the equipment used. Those standards were analyzed to be fitted in a linear model of calibration curve, and two standards were selected as samples: one developed by Lanag at a mole fraction of 380 μmol mol^{-1}, and the other, a CRM from the renowned Dutch Metrology Institute, VSL, at a mole fraction of 300 μmol mol^{-1}, both at the lower limit of the range, in order to determine the limits of quantification of the technique evaluated. Table 5 presents the primary standard mixtures used at the calibration curve fitted for each method and their respective relative standard deviation or coefficient of variation (CV) obtained by each standard mixture.

The linearity was assessed by repeated injections of five primary standards produced at different concentrations englobing the selected range. The linearity is evaluated by the following statistical approach: the coefficient of correlation and the goodness of fit, derived by the part of validation of the response model at ISO 6143. To effectively test the compatibility of a prospective analysis function, calculate the measure of goodness-of-fit (GOF), defined as the maximum value of the weighted differences, $\left|\widehat{xi} - xi\right| / u\left(xi\right)$ and $\left|\widehat{yi} - yi\right| / u\left(yi\right)$, between the coordinates of measured and adjusted calibration points (i=1,2,3…, n). A function is admissible if GOF<2, as well, as r2>0.99.

Results from the primary standards used as samples at the curve of calibration fitted with the responses from the primary standards obtained from each technique can be seen at Table 6. It is presented the relative deviation of the calibrated mole fraction (x_c) obtained by the regression from the gravimetric mole fraction (x_g); the relative expanded uncertainty (U) calculated by the calibration curve fitted; and the number of response repetitions (N) of each day analyzed at each technique evaluated.

The regression analysis was made using the trend line regression tool from Microsoft Excel. And the GOF was obtained by applying the regression based on

Table 5. PSM from calibration curve from CRDS method validation

Primary Standard Mole Fraction (μmol mol^{-1})	Gravimetric Uncertainty (μmol mol^{-1})	Day 1 – CV (%)	Day 2 – CV (%)	Day 3 – CV (%)
370.55	0.08	0.21	0.06	0.05
390.61	0.08	0.23	0.60	0.02
421.32	0.08	0.03	0.16	0.08

Table 6. Results from the sample evaluated at the calibration curve

			CRDS			
Day	**x_g (μmol mol^{-1})**	**x_c (μmol mol^{-1})**	**Δ (%)**	**U (%)**	**CV (%)**	**N**
1	380.027	378.82	0.32	0.42	0.14	05
2	380.027	380.12	-0.02	0.48	0.25	05
3	380.027	380.12	-0.02	0.22	0.21	05

ISO6143:2001, with the calibration curve fitted through the software Xgenline. The linearity results obtained for the coefficient of correlation, r2, and the goodness-of fit (GOF) after applying the regression of the data analyzed are presented on Table 7. In order to verify if the regression from the calibration curve adopted is significant, it was performed an analysis of variance (ANOVA). Another evaluation made from the linear regression of data analysis, was the graph of residuals, which assesses if there is any trend on the results plotted. According to Figure 4, there is no trend observed for CRDS.

Table 7. Results of methods validation

Parameters of Methods Validation Results	CRDS		
	Day 1	*Day 2*	*Day 3*
GOF	1.58	0.23	2.16
r2	0.9991	0.9997	0.9991
Standard error	5.92	3.41	7.67
Observations	43	43	39
F	4783.9	3230.6	32115.0
Fcritic	$4.5\text{x}10^{-18}$	$5.7\text{x}10^{-17}$	$1.9\text{x}10^{-23}$
RSD (%)	0.14	0.25	0.21
s_r (%)	0.20		
s_R (%)	0.53		
ER (%)	0.09		
EN (%)	0.51		
Z-score	0.24		
LOD (μmol mol^{-1})	16		
LOQ (μmol mol^{-1})	50		

Figure 4. Validation parameter: residual trend

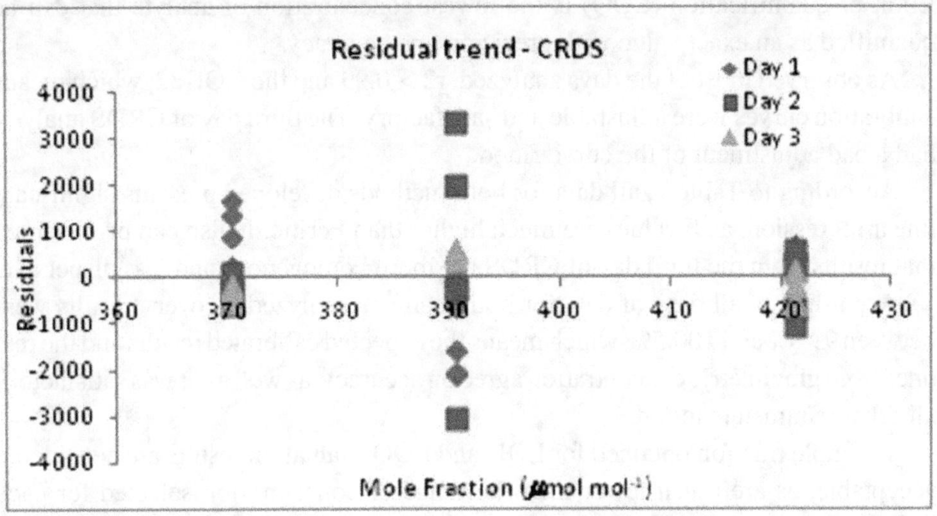

Precision evaluates the dispersion of results between independent and repeated assays. Repeatability is the degree of agreement between the results of successive measurements under the same measurement conditions. Reproducibility is the degree of agreement between the results of successive measurements under varying measurement conditions. Reproducibility was determined internally, only by Inmetro's laboratory, by injecting patterns and samples by different analysts on different days (intermediate precision). The repeatability and the intermediate precision were evaluated by assessing the following statistical approaches: relative standard deviation (RSD %) or coefficient of variation (CV), relative deviation standard for repeatability (sr), relative deviation standard for reproducibility (sR). The acceptance criteria for the parameter quoted before are RSD<1%, sr<2%, and sR<5%. Another parameter evaluated was the repeatability limit (r), which is given by 2.8 times the relative standard deviation of the repeatability (sr), for a significant level of 95%. The reproducibility limit (R) is given by 2.8 times the square root of the relative standard deviation of the reproducibility (sR). Accuracy is the agreement between the result of a test and the reference value accepted as true, and for its evaluation the value of the relative difference between the gravimetric and the analytical concentration derived from the calibration curve (Δ) from the results through the Xgenline software, as it can be seen at on Table 7. Other statistical parameters from accuracy evaluation calculated were: recovery, relative error (ER), normalized error (EN). The Limit of detection (LOD) is the lowest concentration of analyte that can only be detected. Limit of quantification (LOQ) is the lowest concentration of analyte that can be quantified as an exact value with precision and accuracy.

As observed most of the days analyzed, r2 > 0,99 and the GOF<2, which means calibration curves were adjustable and satisfactory. The third day of CRDS analysis had a bad adjustment of the curve fitted.

According to Table 7, all data for both methods developed presents significant linear regression, as F values are much higher than Fcritic. It also can be observed that results from the third day of CRDS are more homogenous and less dispersive.

According to all days of data for both methods analyzed recovery results were between 99.5% and 100.5%, which means the expected calibrated results and the real ones from gravimetric concentration agree on accuracy, as well as, it was satisfactory all other parameters tested.

The mole fraction obtained for LOD and LOQ evaluation results are considered acceptable, as are much lower than the range of concentration selected for each technique.

In summary, most of parameters evaluated presented an accepted acceptance criteria and the method validation for both techniques were considered satisfactory.

Finally, the results obtained on different days were also compared through ANOVA single factor test, which performs the analysis of simple data variance of two or more samples. The acceptance criteria used was the following: if F <Fcritic and P-value> 0.05, the average through different days of the sample are equivalent. This analysis tests the hypothesis that each sample comes from the same baseline probability distribution. Anova single factor results are presented on Table 8.

According to ANOVA single factor criteria, results of average for CRDS are considered satisfactory for both criteria evaluated. All data previously evaluated obtained can be compared to GC results validated from Inmetro's laboratory previous results on the key-comparisons participated among other national metrology institutes for the parameters evaluated (Wessel, 2008; Van der Veen et al, 2011).

Table 8. Anova single factor

<table>
<tr><th>Group</th><th>Counting</th><th>Sum</th><th>Average</th><th>Variance</th><th>F</th><th>Fcritic</th><th>P-value</th></tr>
<tr><td>Day 1</td><td>5</td><td>1977630.6</td><td>395526.1</td><td>297412.1</td><td rowspan="3">54.2</td><td rowspan="3">3.9</td><td rowspan="3">$9.8x10^{-7}$</td></tr>
<tr><td>Day 2</td><td>5</td><td>1995414.1</td><td>399082.8</td><td>997612.9</td></tr>
<tr><td>Day 3</td><td>5</td><td>2003868.3</td><td>400773.7</td><td>689556.7</td></tr>
</table>

When comparing the data given at a certificate of a PRM of CO_2/atmospheric air of 392 μmol/mol from VSL (Dutch NMI) which has a final relative uncertainty of 0.3%, as declared at its certificate (VSL, 2006), it's observed that our average relative uncertainty

is considered at the same level agreement of an experienced NMI standards. The American NMI also develops and puts available standard reference gas mixtures (SRM) of a northern continental air which has the analyte CO_2/atmospheric pure air of 380 μmol/mol with a final relative uncertainty lower than 0.1% (Rhoderick et al., 2016). The low uncertainty acquired at at this SRM is possible to the importance of making isotopic analysis of carbon C_{12}/C_{13} at this type of standard allied to a good system of sample injection that reduces the humidy from atmospheric air during analysis allowing good results on accuracy and precision on the development of this SRM by NIST.

Each analysis uncertainty is then combined with the gravimetry uncertainty to deliver the final expanded uncertainty of the PSM, which can be observed at Table 9 presented.

The value attribution calculation presents the expanded uncertainty (U), and it is expressed as the product of the combined standard uncertainty (u) and the coverage factor $k=2$ (approximately 95% of confidence level) (JCGM, 2008; EURACHEM/CITAC, 2000).

For the final certification of the batch produced, these standard gas mixtures shall go through a short-term and a long-term stability study, to be performed during the period of one year, with periodic analysis to compose the final estimation from this uncertainty contribution of stability of the mixture. It´s important to note that

Table 9. Certified value and expanded uncertainty of PSM of CO2/CDA from Inmetro (Brazil)

PSM	Certified value (mol/mol)
S1	$(370.55 \pm 0.68) \times 10^{-6}$
S2	$(390.60 \pm 0.48) \times 10^{-6}$
S3	$(421.32 \pm 0.88) \times 10^{-6}$
S4	$(453.33 \pm 0.40) \times 10^{-6}$
S5	$(479.94 \pm 0.52) \times 10^{-6}$
S6	$(545.71 \pm 0.78) \times 10^{-6}$
Sample	***$(380.03 \pm 0.50) \times 10^{-6}$***

for gas mixtures standards no homogeneous studies are needed, as the batches are prepared individually, besides considering the homogeneous physical-chemistry behaviour of those component in these gas mixtures.

Precision and accuracy of ambient air measurements also highly depend on the stability of primary and secondary standards. Adsorption or desorption plays an important role in this stability (Miller, Rhoderick, & Guenther, 2014). Besides the choice of the metal, also surface coating as well as humidity are critical for the gas composition (Matsumoto et al., 2005).

INTERNATIONAL COMPARISONS AND INTERCOMPARABILITY

It is task of the NMI to provide the relation to international standards through interlaboratory comparisons with similar foreign institutes. By applying to analysis and/or methods standards which are directly traceable to national primary gases, quality increases, and measurements performed in different laboratories or at different moments are more comparable (RvA-I6.2, October 2004). National Standards Laboratories worldwide, from NMI or from accreditated bodies, carry out comparisons with each other to demonstrate the accuracy and international uniformity of their primary standards. Such comparisons are becoming increasingly important to facilitate growing internation trade (Holland et al., 2001*)*.As a result, the Consultative Committee for Amount of Substance: Chemistry and Biology (CCQM) within the Consultative Committee for Weights and Measures (CIPM) has established formal intercomparison programs whereby a series of key-comparisons of amount of substance measurements most are carried out between selected NMI across the world.

Inmetro PSM of environment area, a standard multicomponent mixture of automotive emission, with carbon monoxide, carbon dioxide and propane compounds,was submitted in 2008 for the first participation of Inmetro in the international key comparison (EURAMET QM-S4) of primary preparation between known NMI and coodinated by VSL under auspices of the Euramet (Van der Veen et al., 2011). This participation aimed to fulfil the necessary preliminary criteria for mutual recognition of measurements, and the results obtained from Inmetro were considered satisfactory as its results were whithin agreement with the reference value result.

In 2003 a comparison CCQM-P41 was carried out between NMIs and WMO laboratories for the determination of greenhouse gases showing a good overall agreement between the participants (van der Veen et al., 2011).

Inmetro has only participated in 2008 in a comparison regarding GHG parameter: the CCQM K-52 (analysis of carbon dioxide in synthetic air) for the level of concentration studied at this paper. The amount–of–substance fraction level of carbon dioxide chosen for this key comparison (360 µmol/mol) represents the ambient level of this component in air (oxygen and nitrogen). The final results of the key-comparison CCQM K-52 are presented in Figure 5 where the relative degrees of equivalence for the laboratories are given relative to the gravimetric value. The uncertainties are, as required by the mutual recognition agreement - MRA (BIPM, 2003), given as 95% confidence intervals. For the evaluation of uncertainty of the degrees of equivalence, the normal distribution has been assumed, and a coverage factor k=2 was used. (Wessel et al., 2008).

Although the high uncertainty from Inmetro measurements, its results agree with the key-comparison relative value (KCRV) within 0.3% relative. In most cases, the deviation from the KCRV is smaller than the associated expanded uncertainty.

Figure 5. Relative degrees of equivalence with uncertainties of CCQM K-52

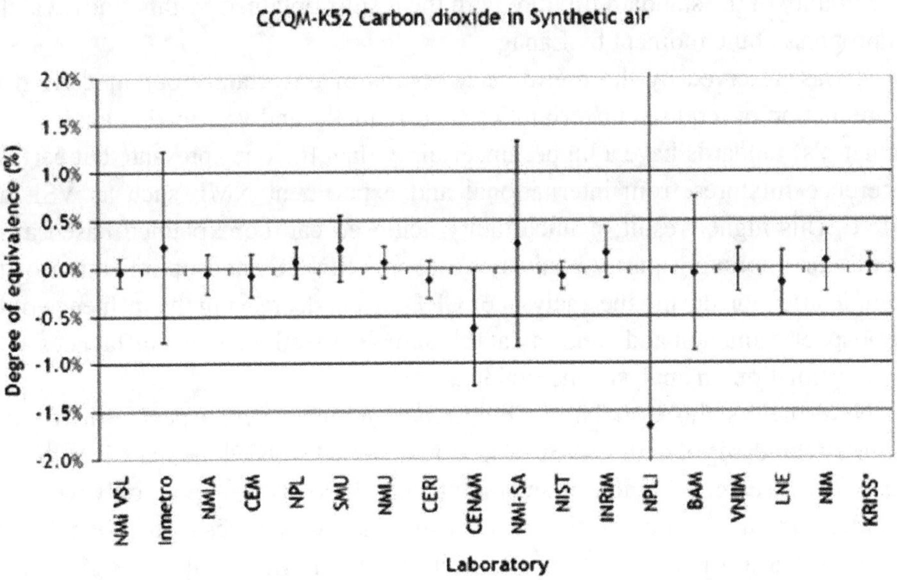

CONCLUSION

Nowadays, the rising concerning on environment regarding global atmospheric air plays an important role in the level of air pollution. So, the analysis of air quality using certified reference material developed by CRM producer and/or by NMI requires reliable and comparable analytical methods in order to reduce the emission of such greenhouse gases. An appropriate CRM can usually be matched to the characteristics of a real sample

This paper describes several aspects regarding new developments in the gravimetric production of primary standard gas mixtures of carbon dioxide in atmospheric synthetic air in Brazil's Metrology Institute, Inmetro, in order to provide traceability and confidence on GHG monitoring research and analysis. Carbon dioxide at atmospheric levels primary standards mixtures were gravimetrically produced according ISO 6142, and .the results were analytically verified by cavity ringdown system with a calibration model regression, according to ISO 6143. The materials were prepared obeying all the gravimetric requirements of ISO Standards, and can be certified after undergoing the requirements of ISO17034. So, in order to finalize the certification process, it's necessary to undergo all batch of PSM prepared and verified through a long term stability study of at least one year to compose the final uncertainty of the standard mixture with the results obtained by this study which is in progress at the moment by Lanag.

It was observed by the preliminary results of uncertainty obtained from the combination of required uncertainties of gravimetry and verification analysis that Inmetro's standards have a higher uncertainty than the ones presented at certified reference mixtures from international and experiment NMI, such as VSL and NIST. This higher result of uncertainty achieved can be explained based at the needs for improving purity analysis using CG-PDHID, and optimization of the sample injection during the analysis of CRDS step, decreasing the influence of the atmospheric ambient and humidity at this sample injection, which influences at the final estimation on analysis uncertainty.

Nevertheless, this work has its importance and innovative aspect as there are no primary standards of carbon dioxide at atmospheric levels with traceability that can be used at research laboratories and monitoring stations that applies only with work standards from gas industry, either with a uncertainty ten times higher than the ones developed at this paper or with no traceability at all. The monitoring laboratories which is part of WMO-GAG network still needs to deal with long time and high cost of importation of standard mixtures, such as the ones from NIST or NOAA, in order to apply traceability and confidence to its measurement results.

We expect this paper will contribute to scientific community to alert to the importance of reliability and traceability in environmental measurements, especially in such a complex field like the monitoring of atmospheric air of GHG parameters, by the use of this kind of reference materials. So, this works presents as novelty and interest to readership the use of primary methods and achievements made in developing of environmental CRM, in order to support the needs of global society and to disseminate traceability and harmonization among countries all over the world measurements of carbon dioxide traces in the atmosphere.

REFERENCES

ABNT. (2012). *ABNT ISO Guia 35:2012 – Materiais de Referência – Princípios gerais e estatísticos para certificação*. Rio de Janeiro: ABNT..

ABNT. (2016). *ABNT ISO Guia 30:2016 – Materiais de Referência – Termos e Definições selecionados*. Rio de Janeiro: ABNT.

ABNT. (2017). *ABNT NBR ISO 17034:2017 – Requisitos gerais para a competência de produtores de materiais de referência*. Rio de Janeiro: ABNT.

Bureau international des poids et mesures (BIPM). (2003). *Annex B to the MRA*. Retrieved August 2017 from http://www.bipm.org/en/cipm-mra/cipm-mra-documents/

Bureau international des poids et mesures - BIPM, IEC, IFCC, ISO, IUPAC, IUPAP, OIML. (1995). *Guide to the expression of uncertainty in measurement*. ISO.

CIPM. (1999). *Mutual recognition of national measurement standards and of calibration and measurement certificates issued by National Metrology Institutes*. CIPM.

CIPM. (1999). *Mutual recognition of national measurement standards and of calibration and measurement certificates issued by national metrology institutes*. Sèvres, France: CIPM.

Custodio, R., Andrade, J. C., & Augusto, F. (1997). O ajuste de funções matemáticas aos dados experimentais. *Quimica Nova*, *20*(2), 219–225. doi:10.1590/S0100-40421997000200016

Dutch Accreditation Council. RvA-I6.2 – Traceability of gases. (2014). Retrieved August 2017 from https://www.rva.nl/en/documents/assessment-documents

EURACHEM / Cooperation on International Traceability in Analytical Chemistry (CITAC). (2000). Guide CG 4, Quantifying uncertainty in analytical measurement (2nd ed.). London: Author.

Fioravante, A. L. (2018). Validation and comparison of calibration techniques for measurements of carbon dioxide in atmospheric air standards. *Journal of Environmental Protection*, *9*, 34–49. doi:10.4236/jep.2018.91004

Fioravante, A. L., Augusto, C. R., Smarçaro da Cunha, V., & d'Avila, L. A. (2017). The challenge of developing primary standards mixtures of carbon dioxide at atmospheric levels to establish traceability to GHG monitoring analysis. *International Journal of Measurement Technologies and Instrumentation Engineering*, *6*(2), 1–15. doi:10.4018/IJMTIE.2017070101

Fioravante, A. L., Guimarães, E. F., Gonzaga, F. B., Ribeiro, C. M., Sobral, S. P., Lopes, J. C., ... Cunha, V. S. (2013). Developments of environmental certified reference material from the Brazilian metrology Institute to support national traceability. *International Journal of Measurement Technologies and Instrumentation Engineering*, *3*(3), 1–17. doi:10.4018/ijmtie.2013070101

Holland, P. E., Woods, P. T., Brookes, C., Uprichard, I., Quincey, P., Robinson, R., ... Alphonso, D. (2001). The development of gas standards and calibration techniques for measurements of vehicle, aircraft and industrial emissions, natural gas, occupational exposure and air quality. *Accreditation and Quality Assurance*, *6*(3), 120–129. doi:10.1007/PL00010447

Inmetro. (2010). *DOC-CGCRE-08 – Orientação sobre validação de métodos analíticos, revisão 03*. Rio de Janeiro: Author.

Inmetro. (2013a). *Dimci 1538/2013, Certificate of CRM 8634.0035: Automotive emissions in nitrogen*. Brazil: Author.

International Conference on Harmonisation (ICH). (1995). *Validation of Analytical Procedures: Methodology, Q2B (CPMP/ICH/281/95)*. Retrieved from http://www.ich.org/LOB/media/MEDIA417.pdf

International Organization for Standardization. (2013). ISO 14065 - Greenhouse gases - Requirements for greenhouse gas validation and verification bodies for use in accreditation or other forms of recognition (2nd ed.). Geneva: ISO.

International Organization for Standardization. (2015). ISO 6142 - Gas analysis Preparation of calibration gas mixtures Gravimetric methods. Geneva: ISO.

ISO. (1994a). ISO 5725-1:1994. Accuracy (trueness and precision) of measurement methods and results -- Part 1: General principles and definitions (and Corrigendum 1:1998). Geneva: ISO.

ISO. (1994b). ISO 5725-2:1994. Accuracy (trueness and precision) of measurement methods and results -- Part 2: Basic method for the determination of repeatability and reproducibility of a standard measurement method (and Technical Corrigendum 1:2002). Geneva: ISO.

ISO. (2001). ISO 6143 - Gas analysis Comparison methods for determining and checking the composition of calibration gas mixtures. Geneva: ISO.

Joint Committee for Guides in Metrology (JCGM). (2008). *Evaluation of measurement data Guide to the expression of uncertainty in measurement*. Retrieved July, 2013, from http://www.bipm.org/utils/common/documents/jcgm/JCGM_100_2008_E.pdf

Leite, F. (2002). *Validação em Análise Química* (4th ed.). Campinas: Editora Átomo.

Matsumoto, N., Watanabe, T., & Kato, K. (2005). Effect of moisture adsorption/desorption on external cylinder surfaces: Influence on gravimetric preparation of reference gas mixtures. *Accreditation and Quality Assurance*, *10*(7), 382–385. doi:10.100700769-005-0002-y

Miller, W. R. Jr, Rhoderick, G. C., & Guenther, F. R. (2014). Investigating adsorption/desorption of carbon dioxide in aluminium compressed gas cylinders. *Analytical Chemistry*, *87*(3), 1957–1962. doi:10.1021/ac504351b PMID:25519817

Milton, M. J. T., & Quinn, T. J. (2001). Primary methods for the measurement of amount of substance. *Metrologia*, *38*(4), 289–296. doi:10.1088/0026-1394/38/4/1

Milton, M. J. T., Vargha, G. M., & Brown, A. S. (2011). Gravimetric methods for the preparation of standard gas mixtures. *Metrologia*, *48*(5), 121–129. doi:10.1088/0026-1394/48/5/R01

National Physical Laboratory - NPL. Xgenline (v1.1) software. (2014). Retrieved August 2017 from http://www.npl.co.uk/science-technology/mathematics-modelling-and-simulation/mathematics-and-modelling-for-metrology/xlgenline-v1.1

Rhoderick, G. C., Kitziz, D., Kelley, M. E., Miller, W. R. Jr, Hall, B., Dlugokencky, E., ... Carney, J. (2016). Development of a Northern Continental Air Standard Reference Material. *Analytical Chemistry J.*, *88*(6), 3376–3385. doi:10.1021/acs.analchem.6b00123 PMID:26890890

Ribani, M., Bottoli, C. G., Collins, C. H., Jardim, I. C. S., & Melo, L. F. C. (2004). Validação em métodos cromatográficos e eletroforéticos. *Quimica Nova*, *27*(5), 771–780. doi:10.1590/S0100-40422004000500017

Tiger Optics, L. L. C. (2011). Prismatic Trace Gas Analyzer Instruction Manual, M8000. Author.

United Nations. (2015). *Framework convention on climate change "Adoption of Paris Agreement"*. Retrieved from https://unfccc.int/resource/docs/2015/cop21/eng/l09r01.pdf

van der Veen, A.M.H., Brinkmann, F.N.C., Arnautovic, M., Besley, L., Heine, H.-J., Lopez Esteban, T., … Dlugokencky, E. (2007a). International comparison CCQM-P41Greenhouse gases. 2. Direct comparison of primary standard gas mixtures. *Metrologia, 44*.

van der Veen, A.M.H., Brinkmann, F.N.C., Arnautovic, M., Besley, L., Heine, H.-J., Lopez Esteban, T., … Dlugokencky, E. (2007b). International comparison CCQM-P41Greenhouse gases. 1. Measurement capability. *Metrologia, 44*.

van der Veen, A.M.H., Oudwater, R., Ziel, P.R., van Wijk, J.I.T., Augusto, C.R., Fioravante, A.L., … Durisova, Z. (2011). EURAMET QM-S4 Final Report – Key comparison automotive gas mixture. *Metrologia, 48*.

van der Veen, A. M. H., Oudwater, R., Ziel, P. R., van Wijk, J. I. T., Augusto, C. R., Fioravante, A. L., ... Durisova, Z. (2011). Supplementary comparison EURAMET. QM-S4: Automotive gas mixtures. *Metrologia, 48*(1A), 08016. doi:10.1088/0026-1394/48/1A/08016

VSL. (2006). *Certificate number 3220285-05 – PRM of Carbon dioxide in synthetic air*. VSL.

Wessel, R.M. (2008). International comparison CCQM-K52: Carbon dioxide in synthetic air. *Metrologia, 45*.

Zuurbier, P., & Vooren, J. V. (2008). *Contributions to climate change mitigation and the environment*. Wageningen Academic Publishers.

Compilation of References

ABNT. (2012). [– *Materiais de Referência – Princípios gerais e estatísticos para certificação*. Rio de Janeiro: ABNT.]. *ABNT ISO Guia*, *35*, 2012.

ABNT. (2012). *ABNT ISO Guia 35:2012 – Materiais de Referência – Princípios gerais e estatísticos para certificação*. Rio de Janeiro: ABNT..

ABNT. (2016). *ABNT ISO Guia 30:2016 – Materiais de Referência – Termos e Definições selecionados*. Rio de Janeiro: ABNT.

ABNT. (2017). [– *Requisitos gerais para a competência de produtores de materiais de referência*. Rio de Janeiro: ABNT.]. *ABNT NBR ISO*, *17034*, 2017.

ABNT. (2017). *ABNT NBR ISO 17034:2017 – Requisitos gerais para a competência de produtores de materiais de referência*. Rio de Janeiro: ABNT.

Acosta, G. G., Verucchi, C. J., & Gelso, E. R. (2004). A current monitoring system for diagnosing electrical failures in induction motors. *Mechanical Systems and Signal Processing*, *20*(1), 953–965.

Affi, N. H., Abdel-Rahman, M. S., & Nassar, A. M. (1998). Effect of alcohol and/or cocaine on blood glutathione and the ultrastructure of the liver of pregnant CF-1 mice. *Toxicology Letters*, *98*(1), 1–12. doi:10.1016/S0378-4274(98)00040-X PMID:9776556

Agência Nacional de Petróleo (ANP). (2011). *Resolução ANP N° 7, Brasília*. Retrieved July, 2013, from http://nxt.anp.gov.br/NXT/gateway.dll/leg/resolucoes_anp/2011/fevereiro/ranp7-2011.xml

Ahmadi-Shokouh, J., & Qiu, R. C. (2009). Ultra-wideband (UWB) communications channel measurements–a tutorial review. *International Journal of Ultra Wideband Communications and Systems*, *1*(1), 11–31. doi:10.1504/IJUWBCS.2009.026447

Al-Aruri, S. D. (1991). *Studies of wavelength-dependent loss effects in optical fiber components and sensors* (MS thesis). Rose-Hulman Institute of Technology. (Report# CASOS-TR-44)

Alaruri, S. D. (2012). Gain, noise figure and efficiency characteristics of an L-band erbium doped fiber amplifier. *International Journal of Measurement Technologies and Instrumentation Engineering*, *2*(4), 1–11. doi:10.4018/ijmtie.2012100101

Alaruri, S. D. (2017). Study of wavelength-dependent bend loss in a step-index multimode fiber-optic microbend displacement sensor. *International Journal of Measurement Technologies and Instrumentation Engineering*, *6*(1), 13–21. doi:10.4018/IJMTIE.2017010102

Alexiou, A., & Haardt, M. (2004). Smart antenna technologies for future wireless systems: Trends and challenges. *IEEE Communications Magazine*, *42*(9), 90–97. doi:10.1109/MCOM.2004.1336725

Ao, M., & Furuichi, S. (2006). *Variable optical attenuator*. US Patent No. 7,065,284 B2.

Arya, V., Murphy, K. A., Wang, A., & Claus, R. O. (1995). Microbend losses in single mode optical fibers: Theoretical and experimental investigation. *Journal of Lightwave Technology*, *13*(10), 1998–2002. doi:10.1109/50.469736

Autschbach, J. (2012). Perspective: Relativistic effect. *The Journal of Chemical Physics*, *136*(1), 1–15. PMID:22519307

Ayhan, B., Trussell, H. J., Chow, M. Y., & Song, M. H. (2008). On the use of a lower sampling rate for broken rotor bar detection with DTFT and AR-based spectrum methods. *IEEE Transactions on Industrial Electronics*, *55*(3), 1421–1434. doi:10.1109/TIE.2007.896522

Bardo, M. T., & Bevins, R. (2000). Conditioned Place Preference: What does it add to our preclinical understanding of drug reward? *Psychopharmacology*, *153*(1), 31–43. doi:10.1007002130000569 PMID:11255927

Bates, C. K. (1988). Medical risks of cocaine use. *The Western Journal of Medicine*, *148*(4), 440. PMID:3291396

Benbouzid, M. E. H. (2000). A Review of induction motors signature analysis as a medium for faults detection. *IEEE Transactions on Industrial Electronics*, *47*(5), 984–993. doi:10.1109/41.873206

Benmeyan, J. C. (1998). Use of interpretation of statistical quality control charts. *International Journal for Quality in Health Care*, *10*(1), 69–73. doi:10.1093/intqhc/10.1.69 PMID:10030790

Bergmann, P. G. (1942). *Introduction to the Theory of Relativity*. Englewood Cliffs, NJ: Prentice-Hall, Inc.

Berthold, J. W. (1995). Historical review of microbend fiber-optic sensors. *Journal of Lightwave Technology*, *13*(7), 1193–1199. doi:10.1109/50.400697

Berthold, J. W., Ghering, W., & Varshneya, D. (1987). Design and characterization of a high temperature fiber-optic pressure transducer. *Journal of Lightwave Technology*, *LT-5*(7), 870–876. doi:10.1109/JLT.1987.1075594

Bhumiwat, S. (2010). Performance of oil and paper in transformer based on IEC 61620 and dielectric response techniques. *IEEE Electrical Insulation Magazine*, *26*(3), 16–23.

Bin, M., & Xinguo, Z. (2010). Study of vehicle weight-in-motion system based on fiber-optic microbend sensor. *IEEE International Conference on Intelligent Computation Technology and Automation*, 458-461. 10.1109/ICICTA.2010.631

Blodt, M., Chabert, M., Regnier, J., & Faucher, J. (2006). Mechanical Load Fault Detection in Induction Motors by Stator Current Time-Frequency Analysis. *IEEE Transactions on Industry Applications*, *42*(6), 1454–1463. doi:10.1109/TIA.2006.882631

Boechat, A. D., Hall, S. D., & Jones, J. (1991). Bend loss in large core multimode optical fiber beam delivery systems. *Applied Optics*, *30*(3), 321–327. doi:10.1364/AO.30.000321 PMID:20581985

Born, M., & Wolf, E. (1999). *Principle of Optics*. Cambridge, UK: Cambridge University Press, seventh expanded edition, Appendix II. doi:10.1017/CBO9781139644181

Boyer, C. S., & Peterson, D. R. (2005). Hepatic biochemical changes as a result of acute cocaine administration in the mouse. *Hepatology (Baltimore, Md.)*, *14*(6), 1209–1216. doi:10.1002/hep.1840140641 PMID:1959871

Buerck, J., & Sensfelder, E. (1999). Optical fiber sensors for the distributed measurement of hydrocarbons. *Proc. SPIE*, 3540, 98-109. 10.1117/12.339785

Bureau international des poids et mesures - BIPM, IEC, IFCC, ISO, IUPAC, IUPAP, OIML. (1995). *Guide to the expression of uncertainty in measurement*. ISO.

Bureau international des poids et mesures (BIPM). (2003). *Annex B to the MRA*. Retrieved August 2017 from http://www.bipm.org/en/cipm-mra/cipm-mra-documents/

Bush, P., Lahti, P., & Werner, R. F. (2014). Colloquium: Quantum root-mean-square error and measurement uncertainty relations. *Reviews of Modern Physics*, *86*(4), 1261–1281. doi:10.1103/RevModPhys.86.1261

Bystrowska, B., Adamczyk, P., Moniczewski, A., Zaniewska, M., Fuxe, K., & Filip, M. (2012). LC/MS/MS evaluation of cocaine and its metabolites in different brain areas, peripheral organs and plasma in cocaine self-administering rats. *Pharmacological Reports*, *64*(1337), 1337–1349. doi:10.1016/S1734-1140(12)70931-3 PMID:23406744

Çalis, H., & Çakir, A. (2007). Rotor bar fault diagnosis in three phase induction motors by monitoring fluctuations of motor current zero crossing instants. *Electric Power Systems Research*, *77*(5), 385–392. doi:10.1016/j.epsr.2006.03.017

Carboni, E., & Vacca, C. (2003). Conditioned place preference. In *Drugs of Abuse* (pp. 481–498). Humana Press.

Casimir, R., Boutleux, E., Clerc, G., & Yahoui, A. (2006). The use of features selection and nearest neighbours rule for faults diagnostic in induction motors. *Engineering Applications of Artificial Intelligence*, *19*(2), 169–177. doi:10.1016/j.engappai.2005.07.004

Center for Substance Abuse Treatment. (1999). Brief Interventions and Brief Therapies for Substance Abuse. Rockville, MD: Substance Abuse and Mental Health Services Administration.

Chen, Z., Lau, D., Teo, J. T., Ng, S. H., Yang, X., & Kei, P. L. (2014). Field test of microbend fiber sensor for hospital use. *Proc. SPIE 9157, 23rd International Conference on Optical Fibre Sensors*, 91579A.

Chen, X., Zhang, Z., Chen, H. H., & Zhang, H. (2015). Enhancing wireless information and power transfer by exploiting multi-antenna techniques. *IEEE Communications Magazine*, *53*(4), 133–141. doi:10.1109/MCOM.2015.7081086

Chen, Z. N., Wu, X. H., Li, H. F., Yang, N., & Chia, M. Y. W. (2004). Considerations for source pulses and antennas in UWB radio systems. *IEEE Transactions on Antennas and Propagation*, *52*(7), 1739–1748. doi:10.1109/TAP.2004.831405

Cherin, A. H. (1983). *An Introduction to Optical Fibers*. New York: McGraw-Hill Book Company.

Chiang, C.N., & Hawks, R.L. (1986). Implications of drug levels in body fluids: Basic concepts. *Urine Testing for Drugs of Abuse, 73*, 62-83.

CIPM. (1999). *Mutual recognition of national measurement standards and of calibration and measurement certificates issued by National Metrology Institutes*. CIPM.

CIPM. (1999). *Mutual recognition of national measurement standards and of calibration and measurement certificates issued by national metrology institutes*. Sèvres, France: CIPM.

Colucci, A. P., Aventaggiato, L., Centrone, M., & Gagliano-Candela, R. (2010). Validation of an Extraction and Gas Chromatography-Mass Spectrometry Quantification Method for Cocaine, Methadone, and Morphine in Postmortem Adipose Tissue. *Journal of Analytical Toxicology, 34*(6), 342–346. doi:10.1093/jat/34.6.342 PMID:20663287

Cordeiro, C., Challapali, K., Birru, D., & Shankar, S. (2005, November). IEEE 802.22: the first worldwide wireless standard based on cognitive radios. In *New Frontiers in Dynamic Spectrum Access Networks, 2005. DySPAN 2005. 2005 First IEEE International Symposium on* (pp. 328-337). IEEE.

Cortez, L. A. B., Stradiotto, N. R., Zanoni, M. A. B., Fraga, I. C. S., & Borges, P. P. (2010). *Sugar cane bioethanol R&D for productivity and sustainability, 4th Part, 21*. São Paulo: Ed.Blücher.

Cunningham, C. L., Gremel, C. M., & Groblewski, P. A. (2006). Drug-induced conditioned place preference and aversion in mice. *Nature Protocols, 1*(4), 1662–1670. doi:10.1038/nprot.2006.279 PMID:17487149

Custodio, R., Andrade, J. C., & Augusto, F. (1997). O ajuste de funções matemáticas aos dados experimentais. *Quimica Nova, 20*(2), 219–225. doi:10.1590/S0100-40421997000200016

DeBoynton, W. L., & Uschitsky, M. (2001). *Fiber coupler variable optical attenuator*. US Patent No. 6,173,106 B1.

Dia, S., Godora, B., Alicalapa, F., & Fabre, A. (2005). Ultra wide-band: State of the art; implementation of a performance-controllable low-noise amplifier. *Turkish Journal of Electrical Engineering and Computer Sciences, 13*(1), 1–23.

Didier, G., Ternisien, E., Caspary, O., & Razik, H. (2007). A new approach to detect broken rotor bars in induction machines by current spectrum analysis. *Mechanical Systems and Signal Processing*, *21*(2), 1127–1142. doi:10.1016/j.ymssp.2006.03.002

Dietrich, C. B., Dietze, K., Nealy, J. R., & Stutzman, W. L. (2001). Spatial, polarization, and pattern diversity for wireless handheld terminals. *IEEE Transactions on Antennas and Propagation*, *49*(9), 1271–1281. doi:10.1109/8.947018

Donlagic, D. (2009). A low bending loss multimode fiber transmission system. *Optics Express*, *17*(24), 22081–22095. doi:10.1364/OE.17.022081 PMID:19997454

Đonlagić, D., & Završnik, M. (1997). Fiber-optic microbend sensor structure. *Applied Optics*, *22*(11), 837–839. PMID:18185679

Du, Y. (1999). Moisture equilibrium in transformer paper-oil system. *IEEE Electrical Insulation Magazine*, *15*(1), 11–20.

Duewer, D. L., Parris, R. M., White, E., May, W. E., & Elbaum, H. (2004). An Approach to the Metrologically Sound Traceable Assessment of the Chemical Purity of Organic Reference Materials. NIST Special Publication 1012.

Dutch Accreditation Council. RvA-I6.2 – Traceability of gases. (2014). Retrieved August 2017 from https://www.rva.nl/en/documents/assessment-documents

Edvard, C. (2011). *A technical report on standard used for transformer oil tests. In 2011 ASTM Standards for Transformer Oil Testing and Complete Analysis of a Transformer's Oil.* Online Electrical Engineering Portal.

Espinoza, L. R., & Alamino, R. P. (2012). Cocaine-induced vasculitis: Clinical and immunological spectrum. *Current Rheumatology Reports*, *14*(6), 532–538. doi:10.100711926-012-0283-1 PMID:22875288

EURACHEM / Cooperation on International Traceability in Analytical Chemistry (CITAC). (2000). Guide CG 4, Quantifying uncertainty in analytical measurement (2nd ed.). London: Author.

EURACHEM / Cooperation on International Traceability in Analytical Chemistry (CITAC). (2000). Guide CG 4. In Quantifying uncertainty in analytical measurement (2nd ed.). London: Author.

Executive Summary from BIOREMA Reference Materials for Biofuel specifications. (2010). Retrieved July, 2013, from http://www.inmetro.gov.br/painelsetorial/palestras/publishable-summary.pdf

Fields, R. E. (1997). Evaluating compliance with FCC guidelines for human exposure to radiofrequency electromagnetic fields. *OET Bulletin, 65*, 10.

Fields, J. N., & Cole, J. H. (1980). Fiber microbend acoustic sensor. *Applied Optics, 19*(3), 265–3267. PMID:20234597

Fields, N. J., Asawa, C. K., Ramer, O. G., & Barnoski, M. K. (1980). Fiber optic pressure sensor. *The Journal of the Acoustical Society of America, 67*(3), 816–818. doi:10.1121/1.383957

Fioravante, A. L. (2018). Validation and comparison of calibration techniques for measurements of carbon dioxide in atmospheric air standards. *Journal of Environmental Protection, 9*, 34–49. doi:10.4236/jep.2018.91004

Fioravante, A. L., Augusto, C. R., Smarçaro da Cunha, V., & d'Avila, L. A. (2017). The challenge of developing primary standards mixtures of carbon dioxide at atmospheric levels to establish traceability to GHG monitoring analysis. *International Journal of Measurement Technologies and Instrumentation Engineering, 6*(2), 1–15. doi:10.4018/IJMTIE.2017070101

Fioravante, A. L., Guimarães, E. F., Gonzaga, F. B., Ribeiro, C. M., Sobral, S. P., Lopes, J. C., ... Cunha, V. S. (2013). Developments of environmental certified reference material from the Brazilian metrology Institute to support national traceability. *International Journal of Measurement Technologies and Instrumentation Engineering, 3*(3), 1–17. doi:10.4018/ijmtie.2013070101

Fraga, I. C. S. F., Ribeiro, C. M., Sobral, S. P., Dias, J. C., Gonçalves, M. A., Borges, P. P., & Gonzaga, F. B. (2012). Certified reference material of bioethanol for metrological traceability in electrochemical parameters analyses. *Talanta, 99*, 99–103. doi:10.1016/j.talanta.2012.05.024 PMID:22967527

Freal, J. B., Zarobila, C. J., & Davis, C. M. (1987). A microbend accelerometer for borehole deployment. *Journal of Lightwave Technology, LT-5*(7), 993–996. doi:10.1109/JLT.1987.1075609

Friedman, Y. (2013). Extending the relativity of time. Journal pf Physics. *Conference Series, 437*(1), 1–23.

Garcia-Rodriguez, O., Secades-Villa, R., Higgins, S. T., Fernandez-Hermida, J. R., Carballo, J. L., Errasti Perez, J. M., & Diaz, S. A. (2009). Effects of voucher-based intervention on abstinence and retention in an outpatient treatment for cocaine addiction: A randomized controlled trial. *Experimental and Clinical Psychopharmacology, 17*(3), 131–138. doi:10.1037/a0015963 PMID:19586227

Gelfand, I. M., & Fomin, S. V. (1963). *Calculus of Variations*. Englewood Cliffs, NJ: Prentice-Hall, Inc.

Ghasemi, A., & Sousa, E. S. (2008). Spectrum sensing in cognitive radio networks: Requirements, challenges and design trade-offs. *IEEE Communications Magazine*, *46*(4), 32–39. doi:10.1109/MCOM.2008.4481338

Goldstein, H. (1959). *Classical Mechanics*. Reading, MA: Addison Wesley Publishing Company, Inc.

Guimarães, E. F., Rego, E. C. P., Cunha, H. C. M., Rodrigues, J. M., & José Daniel Figueroa-Villar, J. D. (2014). Certified Reference Material for Traceability in Environmental Analysis: PAHs in Toluene. *Journal of the Brazilian Chemical Society*, *25*(2), 351–360.

Guimarães, E. F., Vieira, A. A., Rego, E. C. P., Garrido, B. C., Rodrigues, J. M., & Figueroa-Villar, J. D. (2015). Quantitative nuclear magnetic resonance for purity assessment of polycyclic aromatic hydrocarbons. *Metrologia*, *52*(4), L15–L22. doi:10.1088/0026-1394/52/4/L15

Gutman, S. A. (2006, June). Why addiction has a chronic, relapsing course. The neurobiology of addiction: Implications for occupational therapy practice. *Occupational Therapy in Mental Health*, *22*(2), 1–29. doi:10.1300/J004v22n02_01

Hawking, S. W. (1975). Particle creation by black holes. *Communications in Mathematical Physics*, *43*(1), 199–220. doi:10.1007/BF02345020

Hawks, R. L., Kopin, I. J., Colburn, R. W., & Thoa, N. B. (1974). Norcocaine: A pharmacologically active metabolite of cocaine found in brain. *Life Sciences*, *15*(12), 2189–2195. doi:10.1016/0024-3205(74)90035-6 PMID:4621015

Heard, K., Palmer, R., & Zahniser, N. R. (2008). Mechanisms of acute cocaine toxicity. *The Open Pharmacology Journal*, *2*(9), 70–78. doi:10.2174/1874143600802010070 PMID:19568322

Hecht, J. (1999). *Understanding Fiber Optics* (3rd ed.). London: Prentice-Hall International.

Holland, P. E., Woods, P. T., Brookes, C., Uprichard, I., Quincey, P., Robinson, R., ... Alphonso, D. (2001). The development of gas standards and calibration techniques for measurements of vehicle, aircraft and industrial emissions, natural gas, occupational exposure and air quality. *Accreditation and Quality Assurance*, *6*(3), 120–129. doi:10.1007/PL00010447

Howard, E. M. (2012). Causal stability conditions for general relativistic spacetimes. *Journal of Physical Science and Application*, *2*(7), 258–261.

Hu, X. (2007). Cocaine withdrawal and neuro-adaptations in ion channel function. *Molecular Neurobiology*, *35*(1), 95–112. doi:10.1007/BF02700626 PMID:17519508

I.A.R.GRAY. (2010). *A Guide to Transformer Oil Analysis*. I.A.R.GRAY Transformer Chemistry Services. Retrieved from http://www.satcs.co.za/Transformer_Oil_Analysis.pdf

Inmetro. (2010). *Dimci 2354/2010, Certificate of CRM 8315.0001: Ethanol Fuel, Brazil*. Author.

Inmetro. (2010). *DOC-CGCRE-08 – Orientação sobre validação de métodos analíticos, revisão 03*. Rio de Janeiro: Author.

Inmetro. (2012a). *Dimci 1477/2010c, Certificate of CRM 8298.0001: BTEX in methanol, Brazil*. Author.

Inmetro. (2012b). *Dimci 0009/2012, Certificate of CRM 8313.0002: Bioethanol Fuel, Brazil*. Author.

Inmetro. (2013a). *Dimci 1538/2013, Certificate of CRM 8634.0035: Automotive emissions in nitrogen, Brazil*. Author.

Inmetro. (2013a). *Dimci 1538/2013, Certificate of CRM 8634.0035: Automotive emissions in nitrogen*. Brazil: Author.

Inmetro. (2013b). *Dimci 0142/2013, Certificate of CRM 8363.0001a: PAH in toluene, Brazil*. Author.

Inmetro. (2013c). *Dimci 1384/2013, Certificate of CRM 8435.0002: Electrolytic conductivity of 5 $\mu S\ cm^{-1}$, Brazil*. Author.

Inmetro. (2013d). *Dimci0792/2013, Certificate of CRM 8313.0004: Ethanol Fuel, Brazil*. Author.

Inmetro. (2013e). *Dimci 2526/2012, Certificate of CRM 8314.0002a: Ethanol Fuel, Brazil*. Author.

Inmetro. (2015). *Dimci 02715/2014a, Certificate of CRM 8363.0001b: PAH in toluene, Brazil*. Author.

Inmetro. (2016a). *Dimci 1477/2010g, Certificate of CRM 8298.0001: BTEX in methanol, Brazil*. Author.

Inmetro. (2016b). *Dimci 2526/2012c, Certificate of CRM 8314.0002a: Ethanol Fuel, Brazil*. Author.

Interlaboratory Comparison in Anhydrous Bioethanol. (n.d.). Retrieved July, 2013, from http://www.inmetro.gov.br/metcientifica/pdf/Protocol-Bioethanol-PEAAL.pdf

International Agency for Research on Cancer (IARC). (2010). *Complete list of agents, mixtures and Exposures Evaluated and Their Classification*. Retrieved July, 2013, from http://www.iarc.fr

International Conference on Harmonisation (ICH). (1995). *Validation of Analytical Procedures: Methodology, Q2B (CPMP/ICH/281/95)*. Retrieved from http://www.ich.org/LOB/media/MEDIA417.pdf

International Organization for Standardization. (2001). ISO 6143 - Gas analysis Comparison methods for determining and checking the composition of calibration gas mixtures. Geneva: ISO.

International Organization for Standardization. (2013). ISO 14065 - Greenhouse gases - Requirements for greenhouse gas validation and verification bodies for use in accreditation or other forms of recognition (2nd ed.). Geneva: ISO.

International Organization for Standardization. (2015). ISO 6142 - Gas analysis Preparation of calibration gas mixtures Gravimetric methods. Geneva: ISO.

ISO. (1994a). ISO 5725-1:1994. Accuracy (trueness and precision) of measurement methods and results -- Part 1: General principles and definitions (and Corrigendum 1:1998). Geneva: ISO.

ISO. (1994b). ISO 5725-2:1994. Accuracy (trueness and precision) of measurement methods and results -- Part 2: Basic method for the determination of repeatability and reproducibility of a standard measurement method (and Technical Corrigendum 1:2002). Geneva: ISO.

ISO. (2001). ISO 6143 - Gas analysis Comparison methods for determining and checking the composition of calibration gas mixtures. Geneva: ISO.

James, R. C., Roberts, S. M., & Harbison, R. D. (1983). The perturbation of hepatic glutathione by α-adrenergic agonists. *Fundamental and Applied Toxicology*, *3*(4), 303–308. doi:10.1016/S0272-0590(83)80144-4 PMID:6138290

Jensen, H. D. (2009). *Final Report on CCQM-K36.1*. Metrologia, 46, Tech. Suppl., 08004.

Jensen, M. A., & Wallace, J. W. (2004). A review of antennas and propagation for MIMO wireless communications. *IEEE Transactions on Antennas and Propagation*, *52*(11), 2810–2824. doi:10.1109/TAP.2004.835272

Joint Committee for Guides in Metrology (JCGM). (2008). *Evaluation of measurement data Guide to the expression of uncertainty in measurement*. Retrieved July, 2013, from http://www.bipm.org/utils/common/documents/jcgm/JCGM_100_2008_E.pdf

Kaila, V. R. I., & Annila, A. (2008). Natural selection for least action. *Proceeding of The Royal Society A*, *464*(1), 3055–3070. doi:10.1098/rspa.2008.0178

Kanel, G. C., Cassidy, W., Shuster, L., & Reynolds, T. B. (1989). Cocaine-induced liver cell injury: Comparison of morphological features in man and in experimental models. *Hepatology (Baltimore, Md.)*, *11*(4), 646–651. doi:10.1002/hep.1840110418 PMID:2328957

Kao, C. K. (1982). *Optical fiber systems: Technology*. New York: Design and Applications, McGraw-Hill Company.

Kennedy, A. P., Gross, R. E., Whitfield, N., Drexler, K. P. G., & Kilts, C. D. (2012). A controlled trial of the adjunct use of D-cycloserine to facilitate cognitive behavioral therapy outcomes in a cocaine-dependent population. *Addictive Behaviors*, *37*(8), 900–907. doi:10.1016/j.addbeh.2012.03.008 PMID:22578380

Khaleel, H. R., Al-Rizzo, H. M., Rucker, D. G., Rahmatallah, Y. A., & Mohan, S. (2011, July). Mutual coupling reduction of dual-band printed monopoles using MNG metamaterial. In *Antennas and Propagation (APSURSI), 2011 IEEE International Symposium on* (pp. 2219-2222). IEEE. 10.1109/APS.2011.5996956

Kildal, P. S., & Rosengren, K. (2004). Correlation and capacity of MIMO systems and mutual coupling, radiation efficiency, and diversity gain of their antennas: Simulations and measurements in a reverberation chamber. *IEEE Communications Magazine*, *42*(12), 104–112. doi:10.1109/MCOM.2004.1367562

Koob, G. F., & Le Moal, M. (2001). Drug addiction, dysregulation of reward, and allostasis. *Neuropsychopharmacology*, *24*(2), 97–129. doi:10.1016/S0893-133X(00)00195-0 PMID:11120394

Krohn, D., MacDougall, T., & Mendez, A. (2014). *Fiber optic sensors fundamentals and applications* (4th ed.). Bellingham, WA: SPIE Press. doi:10.1117/3.1002910

Lagakos, N., & Bucaro, J. A. (1987). Optimizing fiber optic microbend sensor. *Proceedings of the Society for Photo-Instrumentation Engineers*, *718*, 12–20. doi:10.1117/12.937494

Lagakos, N., Cole, J. H., & Bucaro, J. A. (1987). Microbend fiber-optic sensor. *Applied Optics*, *26*(11), 2171–2180. doi:10.1364/AO.26.002171 PMID:20489839

Lagakos, N., Litovitz, T., Macedo, P., Mohr, R., & Meister, R. (1981). Multimode optical fiber displacement sensor. *Applied Optics*, *20*(2), 167–168. doi:10.1364/AO.20.000167 PMID:20309083

Lagakos, N., Trott, W., Hickman, T., Cole, J., & Bucaro, J. (1989). Microbend fiber optic sensor as extended hydrophone. *IEEE Journal of Quantum Electronics*, *OF-18*, 1633–1638.

Landau, L. D., & Lifshits, E. M. (1959). *The Classical Theory of Fields*. Reading, MA: Addison Wesley Publishing Company, Inc.

Larsson, E. G., Edfors, O., Tufvesson, F., & Marzetta, T. L. (2014). Massive MIMO for next generation wireless systems. *IEEE Communications Magazine*, *52*(2), 186–195. doi:10.1109/MCOM.2014.6736761

Lee, J. M., Kim, K. B., Ryu, H. K., & Woo, J. M. (2012). A compact ultrawideband MIMO antenna with WLAN band-rejected operation for mobile devices. *IEEE Antennas and Wireless Propagation Letters*, *11*, 990–993. doi:10.1109/LAWP.2012.2214431

Lehpamer, H. (2016). *Introduction to power utility communications*. Artech House.

Leite, F. (2002). *Validação em Análise Química* (4th ed.). Campinas: Editora Átomo.

Levine, S. R., Brust, J. C. M., Futrell, N., Ho, K.-L., Blake, D., Millikan, C. H., ... Welch, K. M. A. (1990). Cerebrovascular complications of the use of the Crack form of alkaloidal cocaine. *The New England Journal of Medicine*, *323*(11), 699–704. doi:10.1056/NEJM199009133231102 PMID:2388668

Li, X., Chena, Y., Dengb, Y., Lia, X., & Lia, J. (2009). Novel microbending optical fiber temperature sensor. *Proc. of SPIE*, 7508, 75080N-1 to 75080N-5. 10.1117/12.839634

Li, J. F., Chu, Q. X., & Huang, T. G. (2012). A compact wideband MIMO antenna with two novel bent slits. *IEEE Transactions on Antennas and Propagation*, *60*(2), 482–489. doi:10.1109/TAP.2011.2173452

Li, J. F., Chu, Q. X., Li, Z. H., & Xia, X. X. (2013). Compact dual band-notched UWB MIMO antenna with high isolation. *IEEE Transactions on Antennas and Propagation*, *61*(9), 4759–4766. doi:10.1109/TAP.2013.2267653

Lumholt, O., Bjarklev, A., Dahl-Petersen, S., Larsen, C. C., Povlsen, J. H., Rasmussen, T., & Rottwitt, K. (1991). Simple fiber optic low-temperature sensor that uses micro-bending loss. *Optics Letters*, *16*(17), 1355–1357. doi:10.1364/OL.16.001355 PMID:19776969

MacLachlan, P. L., & Hodge, H. C. (1938). The Influence of Cocaine Feeding on the Liver Lipids of the White Mouse. *The Journal of Biological Chemistry*, *104*, 721–726.

Mahajan, M. S. (2001). *Statistical quality control* (3rd ed.). New York, NY: McGraw-Hill.

Maldonando, R., & Rodriguez de Fonseca, F. (2002). Cannabinoid Addiction: Behavioural Models and Neural Correlates. *The Journal of Neuroscience*, *22*(9), 3326–3331. doi:10.1523/JNEUROSCI.22-09-03326.2002 PMID:11978807

Mandavgade, N.K., Jaju, S.B., & Lakhe, R.R. (2012). Identifying uncertainty sources in measurements – a methodology. *Journal of Measurement Science and Instrumentation*, *3*(4), 307–312.

Mandavgade, N. K., Jaju, S. B., & Lakhe, R. R. (2011a). Determination of uncertainty in gross calorific value of coal using Bomb calorimeter. *International Journal of Measurement Technologies and Instrumentation*, *1*(4), 45–52. doi:10.4018/ijmtie.2011100104

Mandavgade, N. K., Jaju, S. B., & Lakhe, R. R. (2011b). Mathematical modeling of effects of various factors on uncertainty of mea-surement in material testing. *Proceedings of the International Conference on Mechanical Engineer-ing and Technology*.

Mandavgade, N. K., Jaju, S. B., & Lakhe, R. R. (2012). Evaluation of uncertainty of measurement of automatic tan delta and resistivity test set for transformer oil. *Int. J. Metrol. Qual. Eng.*, *3*(1), 39–45. doi:10.1051/ijmqe/2012004

Marcuse, D. (1976a). Curvature loss formula for optical fibers. *Journal of the Optical Society of America*, *66*(3), 216–220. doi:10.1364/JOSA.66.000216

Marcuse, D. (1976b). Field deformation and loss caused by curvature of optical fibers. *Journal of the Optical Society of America*, *66*(4), 311–320. doi:10.1364/JOSA.66.000311

Martin, M., Ledent, C., Parmentier, M., Monaldonado, R., & Valverde, O. (2000). Cocaine, but not morphine, induces conditioned place preference and sensitization to locomotor responses in CB1 knockout mice. *The European Journal of Neuroscience*, *12*(11), 4038–4046. doi:10.1046/j.1460-9568.2000.00287.x PMID:11069600

Massaro, A. (2011). *Theory, Modeling, Technology and applications of Micro/Nano quantum electronic and photonic devices*. Kerala, India: Transworld Research Network.

Massaro, A., & Massaro, P. A. (2011). An observation on least action principle in classical mechanics oriented on the evaluation of relativistic error concerning the measurements of light propagating in a liquid. *International Journal of Measurement Technologies and Instrumentation Engineering*, *1*(3), 14–24. doi:10.4018/ijmtie.2011070102

Matias, I. R., Ikezawa, S., & Corres, J. (2017). Fiber Optic Sensors: current status and future possibilities. Springer.

MATLAB. (2015). *Primer, R2015b*. The Math Works, Inc. Retrieved from http://www.mathworks.com

Matsumoto, N., Watanabe, T., & Kato, K. (2005). Effect of moisture adsorption/desorption on external cylinder surfaces: Influence on gravimetric preparation of reference gas mixtures. *Accreditation and Quality Assurance*, *10*(7), 382–385. doi:10.100700769-005-0002-y

Maude-Griffin, P. M., Hohenstein, J. M., Humfleet, G. L., Reilly, P. M., Tusel, D. J., & Hall, S. M. (1998). Superior efficacy of cognitive-behavioral therapy for urban crack cocaine abusers: Main and matching effects. *Journal of Consulting and Clinical Psychology*, *66*(5), 832–837. doi:10.1037/0022-006X.66.5.832 PMID:9803702

Mazhar, W., Tarar, M. A., Tahir, F. A., Ullah, S., & Bhatti, F. A. (2013). Compact microstrip patch antenna for ultra-wideband applications. *PIERS Proceedings*, 1100-1104.

Miller, W. R. Jr, Rhoderick, G. C., & Guenther, F. R. (2014). Investigating adsorption/desorption of carbon dioxide in aluminium compressed gas cylinders. *Analytical Chemistry*, *87*(3), 1957–1962. doi:10.1021/ac504351b PMID:25519817

Milton, M. J. T., & Quinn, T. J. (2001). Primary methods for the measurement of amount of substance. *Metrologia*, *38*(4), 289–296. doi:10.1088/0026-1394/38/4/1

Milton, M. J. T., Vargha, G. M., & Brown, A. S. (2011). Gravimetric methods for the preparation of standard gas mixtures. *Metrologia*, *48*(5), 121–129. doi:10.1088/0026-1394/48/5/R01

Misra, A. L., Nayak, P. K., Bloch, R., & Mule, S. J. (1975). Estimation and disposition of [3H] benzoylecgonine and pharmacological activity of some cocaine metabolites. *The Journal of Pharmacy and Pharmacology*, *27*(10), 784–786. doi:10.1111/j.2042-7158.1975.tb09404.x PMID:241797

Misra, A. L., & Pontani, R. B. (1977). Disposition and metabolism of psi-cocaine (dextrococaine) in the rat. *Drug Metabolism and Disposition: the Biological Fate of Chemicals*, *5*(6), 556–563. PMID:21779

Najam, A. I., Duroc, Y., & Tedjini, S. (2010, April). Design and analysis of MIMO antennas for UWB communications. In *Antennas and Propagation (EuCAP), 2010 Proceedings of the Fourth European Conference on* (pp. 1-5). IEEE.

Narasimhachar, V., Poostindouz, A., & Gour, G. (2016). Uncertainty, joint uncertainty, and the quantum uncertainty principle. *New Journal of Physics*, *18*(1), 1–11.

National Institute on Drug Abuse. (2008). *Cocaine Drug Facts*. Retrieved from http://www.nida.nih.gov/pdf/infofacts/Cocaine08.pdf

National Physical Laboratory - NPL. Xgenline (v1.1) software. (2014). Retrieved August 2017 from http://www.npl.co.uk/science-technology/mathematics-modelling-and-simulation/mathematics-and-modelling-for-metrology/xlgenline-v1.1

Neves, L. A., Almeida, R. R., Rego, E. P., Rodrigues, J. M., Carvalho, L. J., & Goulart, A. L. M. (2015). Certified reference material of volatile organic compounds for environmental analysis: BTEX in methanol. *Analytical and Bioanalytical Chemistry*, *407*, 3225–3229. doi:10.100700216-014-8314-6 PMID:25450051

Norton, J. D. (1993). General covariance and the foundations of general relativity: Eight decades of dispute. *Reports on Progress in Physics*, *56*(1), 791–858. doi:10.1088/0034-4885/56/7/001

Nyambuya, G. G., & Simango, W. (2014). On the gravitational bending of light. *International Journal of Astronomy and Astrophysics*, *4*(1), 250–263. doi:10.4236/ijaa.2014.41022

Oliveira, A. R. E. (2014). History of Two Fundamental Principles of Physics: Least Action and Conservation of Energy. *Advances in Historical Studies*, *3*(1), 83–92. doi:10.4236/ahs.2014.32008

Park, J. (2011). Plastic optical fiber sensor for measuring driver-gripping force. *Optical Engineering Letters, 50*(2), 02050-1 - 02050-3.

Pasanen, M., Pellinen, P., Stenback, F., Juvonen, R. O., Raunio, H., & Pelkonen, O. (1995). The role of CYP enzymes in cocaine-induced liver damage. *Archives of Toxicology*, *69*(5), 287–290. doi:10.1007002040050172 PMID:7654131

Paschotta, R. (2010). *Optical Fiber technology, SPIE Field Guides, Vol. FG16*. Bellingham, WA: SPIE Press. doi:10.1117/3.853722

Porcino, D., & Hirt, W. (2003). Ultra-wideband radio technology: Potential and challenges ahead. *IEEE Communications Magazine*, *41*(7), 66–74. doi:10.1109/MCOM.2003.1215641

Prasanna, K. M., & Behera, S. K. (2013, April). Compact two-port UWB MIMO antenna system with high isolation using a fork-shaped structure. In *Communications and Signal Processing (ICCSP), 2013 International Conference on* (pp. 726-729). IEEE. 10.1109/iccsp.2013.6577151

Rappaport, T. S., Annamalai, A., Buehrer, R. M., & Tranter, W. H. (2002). Wireless communications: Past events and a future perspective. *IEEE Communications Magazine*, *40*(5), 148–161. doi:10.1109/MCOM.2002.1006984

Rebec, G. V. (2012). *Cocaine and Amphetamines in:eLS*. Chichester, UK: John Wiley & Sons Ltd. doi:10.1002/9780470015902.a0000042.pub3

Register, U. D., & Bartlett, R. G. (1954). Relationship of adrenalin to tissue sulfhydryl compounds. *Science*, *120*(3107), 109–110. doi:10.1126cience.120.3107.109 PMID:13178662

Remouche, M., Georges, F., & Meyrueis, P. (2013). Stress sensing by an optical fiber sensor: Method and process for the characterization of the sensor response depending on several designs. *Optics and Photonics Journal*, *3*(02), 194–203. doi:10.4236/opj.2013.32032

Rhoderick, G. C., Kitziz, D., Kelley, M. E., Miller, W. R. Jr, Hall, B., Dlugokencky, E., ... Carney, J. (2016). Development of a Northern Continental Air Standard Reference Material. *Analytical Chemistry J.*, *88*(6), 3376–3385. doi:10.1021/acs.analchem.6b00123 PMID:26890890

Ribani, M., Bottoli, C. G., Collins, C. H., Jardim, I. C. S., & Melo, L. F. C. (2004). Validação em métodos cromatográficos e eletroforéticos. *Quimica Nova, 27*(5), 771–780. doi:10.1590/S0100-40422004000500017

Ribeiro, L. A., Rosolem, J. B., Dini, D. C., Floridia, C., Hortencio, C. A., da Costa, E. F., . . . Durelli, A. S. (2011). Fiber optic bending loss sensor for application on monitoring of embankment dams, Microwave & Optoelectronics Conference (IMOC), 2011 SBMO/IEEE MTT-S International, 637-641.

Robinson, T.E. & Berridge, K.C. (2000). The psychology and neurobiology of addiction: an incentive-sensitization view. *Addiction, 95*(8s2), 91-117.

Robinson, J., Whelan, C. D., & Kent Haggerty, N. (2004). Trends in advanced motor protection and monitoring. *IEEE Transactions on Industry Applications, 40*(3), 853–860. doi:10.1109/TIA.2004.827472

Rodrigues, J. M., Guimarães, E. F., Sousa, M. V. B., Silva, V. F., Fraga, I. C. S., Souza, V., & Cunha, V. S. (2012). Certified Reference Material to water content determination in bioethanol fuel. *Quimica Nova, 35*(5), 1011–1015. doi:10.1590/S0100-40422012000500026

Rook, E. J., Huitema, A. D. R., Mv Ree, J., & Beijnen, J. H. (2006). Pharmacokinetics and pharmacokinetic variability of heroin and its metabolites: Review of the literature. *Current Clinical Pharmacology, 1*(1), 109–118. doi:10.2174/157488406775268219 PMID:18666382

Roriz, P., Frazao, O., Lobo-Ribeiro, A.B., Santos, J.L. & Simoes, J.A. (2013). Review of fiber-optic pressure sensors for biomedical applications and biomechanical applications. *Journal of Biomedical Optics, 18*(5), 050903-1 to 050903-18.

Schermer, R. T., & Cole, J. H. (2007). Improved bend loss formula verified for optical fiber by simulation and experiment. *IEEE Journal of Quantum Electronics, 42*(10), 899-909.

See, C. H., Abd-Alhameed, R. A., Abidin, Z. Z., McEwan, N. J., & Excell, P. S. (2012). Wideband printed MIMO/diversity monopole antenna for WiFi/WiMAX applications. *IEEE Transactions on Antennas and Propagation, 60*(4), 2028–2035. doi:10.1109/TAP.2012.2186247

Sen, D. (2014). The uncertainty relations in quantum mechanics. *Current Science, 107*(2), 213–218.

Serway, R. A., Moses, C. J., & Moyer, C. A. (2005). *Modern Physics*. Belmont, WA: Thomson Learning, Inc.

Shuster, L., Eleanor, C., & Sudhakar, S. W. (1983). Metabolism of cocaine and norcocaine to N-hydroxynorcocaine. *Biochemical Pharmacology*, *32*(20), 3045–3051. doi:10.1016/0006-2952(83)90247-2 PMID:6639673

Sibille, A., Oestges, C., & Zanella, A. (2010). *MIMO: from theory to implementation*. Academic Press.

Snyder, W., White, I., & Mitchell, D. J. (1975). Radiation from bent optical waveguides. *Electronics Letters*, *11*(15), 332–333. doi:10.1049/el:19750254

Soh, P. J., Vandenbosch, G. A., Mercuri, M., & Schreurs, D. M. P. (2015). Wearable wireless health monitoring: Current developments, challenges, and future trends. *IEEE Microwave Magazine*, *16*(4), 55–70. doi:10.1109/MMM.2015.2394021

Solovej, J. P. (2011). The stability of matter in quantum mechanics. *Bulletin of the American Mathematical Society*, *50*(1), 169–174. doi:10.1090/S0273-0979-2011-01366-0

Sora, I., Wichems, C., Takahashi, N., Li, X., Zeng, Z., Revay, R., ... Uhl, G. R. (1998). Cocaine reward models: Conditioned place preference can be established in dopamine- and in serotonin-transporter knockout mice. *Proceedings of the National Academy of Sciences of the United States of America*, *95*(13), 7699–7704. doi:10.1073/pnas.95.13.7699 PMID:9636213

Spillman, W. B., & Udd, E. (2014). Field guide to fiber optic sensors (vol. FG34). SPIE Press.

Stannett, A. W. (1951). The conductivity of hydrocarbon transformer oil containing water and solid conducting particles. *British Journal of Applied Physics*, *2*(4), 304. doi:10.1088/0508-3443/2/4/304

Stevens, P. (2008). The Technical Report on 'A Methodology for Assessing the Performance of National Oil Companies'. Academic Press.

Styer, D. F. (1995). Common misconceptions regarding quantum mechanics. *American Journal of Physics*, *64*(1), 31–34. doi:10.1119/1.18288

Subhendu, D. (2013). Assumptions in quantum mechanics. *International Journal and Mathematical Physics*, *3*(2), 53–68.

Subramanian, P. (2016). *Improving reliability aware flooding algorithms in wireless multi-hop sensor networks*. Academic Press.

Svetlichny, G. (1997). On Relativistic Non-linear Quantum Mechanics. *Symmetry in Nonlinear Mathematical Physics*, *2*(1), 262–269.

Thakur, A., Wadhwani, S., & Sondhiya, V. (2013). Health monitoring of rotating electrical machine using soft computing techniques: A Review. *International Journal of Scientific and Research Publications*, *3*(11), 1–3.

Thompson, M. L., Shuster, L., & Shaw, K. (1979). Cocaine-induced hepatic necrosis in mice—The role of cocaine metabolism. *Biochemical Pharmacology*, *28*(15), 2389–2395. doi:10.1016/0006-2952(79)90705-6 PMID:497018

Tiger Optics, L. L. C. (2011). Prismatic Trace Gas Analyzer Instruction Manual, M8000. Author.

Tsai, L. and Zager, D. (2005). Medical Treatment in autism. *Autism spectrum disorders: Identification, education and treatment*, 395-492.

Tzschentke, T. M. (1998). Measuring reward with the conditioned place preference paradigm: A comprehensive review of drug effects, recent progress and new issues. *Progress in Neurobiology*, *56*(6), 613–672. doi:10.1016/S0301-0082(98)00060-4 PMID:9871940

Udd, E., & Spillman, W. B. (2011). *Fiber optic sensors: an introduction for engineers and scientists* (2nd ed.). John Wiley & Sons, Inc. doi:10.1002/9781118014103

United Nations. (2015). *Framework convention on climate change "Adoption of Paris Agreement"*. Retrieved from https://unfccc.int/resource/docs/2015/cop21/eng/l09r01.pdf

UNODC. (2010). *United Nations Office on Drugs and Crime. World Drug Report 2010*. United Nations Publication.

van der Veen, A.M.H., Brinkmann, F.N.C., Arnautovic, M., Besley, L., Heine, H.-J., Lopez Esteban, T., … Dlugokencky, E. (2007a). International comparison CCQM-P41Greenhouse gases. 2. Direct comparison of primary standard gas mixtures. *Metrologia, 44*.

van der Veen, A.M.H., Brinkmann, F.N.C., Arnautovic, M., Besley, L., Heine, H.-J., Lopez Esteban, T., … Dlugokencky, E. (2007b). International comparison CCQM-P41Greenhouse gases. 1. Measurement capability. *Metrologia, 44*.

van der Veen, A.M.H., Oudwater, R., Ziel, P.R., van Wijk, J.I.T., Augusto, C.R., Fioravante, A.L., … Durisova, Z. (2011). EURAMET QM-S4 Final Report – Key comparison automotive gas mixture. *Metrologia, 48*.

van der Veen, A. M. H., Oudwater, R., Ziel, P. R., van Wijk, J. I. T., Augusto, C. R., Fioravante, A. L., ... Durisova, Z. (2011). Supplementary comparison EURAMET. QM-S4: Automotive gas mixtures. *Metrologia*, *48*(1A), 08016. doi:10.1088/0026-1394/48/1A/08016

Vaughan, R. G., & Andersen, J. B. (1987). Antenna diversity in mobile communications. *IEEE Transactions on Vehicular Technology*, *36*(4), 149–172. doi:10.1109/T-VT.1987.24115

Volkow, N.D. (2009). What Is Cocaine? In *Cocaine: Abuse and Addiction*. NIDA Research Report Series 3-7.

VSL. (2006). *Certificate number 3220285-05 – PRM of Carbon dioxide in synthetic air*. VSL.

Weiss, J. D. (1989). Fiber optic strain gage. *Journal of Lightwave Technology*, *7*(9), 1308–1318. doi:10.1109/50.50709

Weiss, R. D., Ostacher, M. J., Otto, M. W., Calabrese, J. R., Fossey, M., Wisniewski, S. R., ... Sachs, G. S. (2005). Does recovery from substance use disorder matter in patients with bipolar disorder? *The Journal of Clinical Psychiatry*, *66*(6), 730–735. doi:10.4088/JCP.v66n0609 PMID:15960566

Werbach, K. (2002). The New Wireless Paradigm. *Spectrum Series Working Paper, 6*.

Wessel, R.M. (2008). International comparison CCQM-K52: Carbon dioxide in synthetic air. *Metrologia, 45*.

Westgard, J. O., Barry, P. L., & Hunl, M. R. (1981). A multi rule shewart chart for quality control in clinical laboratory. *Clinical Chemistry*, *27*(3), 493–501. PMID:7471403

Westgard, J. O., Barry, P. L., & Hunl, M. R. (1981). A multi rule shewart chart for quality control in clini-cal laboratory. *Clinical Chemistry*, *27*(3), 493–501. PMID:7471403

Williams, J., & Wasserberger, J. (2006). Crack cocaine causing fatal vasoconstriction of the aorta. *The Journal of Emergency Medicine*, *31*(2), 181–184. doi:10.1016/j.jemermed.2005.08.021 PMID:17044582

Xiao, Y., J., Naihuan J., & Fei S. M. (2016). Weighted Uncertainty Relations. *Nature Scientific Reports 6*(23201).

Xinzhong, W. (2016). Analytical mechanics structure of quantum mechanics and relativity space-time. *SCIREA Journal of Physics*, *1*(2), 108–124.

Xiong, T. P., Yan, L. L., Ma, Z. H., Zhou, F., Chen, L., Yang, W. L., ... Busch, P. (2017). Optimal joint measurements of complementary observables by a single trapped ion. *New Journal of Physics*, *19*(1), 1–9.

Yang, X., Chen, Z., Elvin, C. S. M., Janice, L. H. Y., Ng, S. H., Teo, J. T., & Wu, R. (2015). Textile fiber optic microbend sensor used for heartbeat and respiration monitoring. *IEEE Sensors Journal*, *15*(2), 757–761. doi:10.1109/JSEN.2014.2353640

Yang, X., Liu, Y., Xu, Y. X., & Gong, S. X. (2017). Isolation Enhancement in Patch Antenna Array With Fractal UC-EBG Structure and Cross Slot. *IEEE Antennas and Wireless Propagation Letters*, *16*, 2175–2178. doi:10.1109/LAWP.2017.2703170

Yao, S. K., & Asawa, C. K. (1983). Microbending fiber optic sensing. *Proceedings of the Society for Photo-Instrumentation Engineers*, *412*, 9–12. doi:10.1117/12.935788

Yuhua, F., & Fengjuan, Z. (2013). *Unsolved Problems in Special and General Relativity*. Columbus, OH: Education Publishing.

Zhang, S., Lau, B. K., Sunesson, A., & He, S. (2012). Closely-packed UWB MIMO/diversity antenna with different patterns and polarizations for USB dongle applications. *IEEE Transactions on Antennas and Propagation*, *60*(9), 4372–4380. doi:10.1109/TAP.2012.2207049

Zhang, S., & Pedersen, G. F. (2016). Mutual coupling reduction for UWB MIMO antennas with a wideband neutralization line. *IEEE Antennas and Wireless Propagation Letters*, *15*, 166–169. doi:10.1109/LAWP.2015.2435992

Zuurbier, P., & Vooren, J. V. (2008). *Contributions to climate change mitigation and the environment*. Wageningen Academic Publishers.

Related References

To continue our tradition of advancing academic research, we have compiled a list of recommended IGI Global readings. These references will provide additional information and guidance to further enrich your knowledge and assist you with your own research and future publications.

Abed, S., Khir, T., & Ben Brahim, A. (2016). Thermodynamic and Energy Study of a Regenerator in Gas Turbine Cycle and Optimization of Performances. *International Journal of Energy Optimization and Engineering*, *5*(2), 25–44. doi:10.4018/IJEOE.2016040102

Abu Bakar, W. A., Abdullah, W. N., Ali, R., & Mokhtar, W. N. (2016). Polymolybdate Supported Nano Catalyst for Desulfurization of Diesel. In T. Saleh (Ed.), *Applying Nanotechnology to the Desulfurization Process in Petroleum Engineering* (pp. 263–280). Hershey, PA: IGI Global. doi:10.4018/978-1-4666-9545-0.ch009

Addo-Tenkorang, R., Helo, P., & Kantola, J. (2016). Engineer-To-Order Product Development: A Communication Network Analysis for Supply-Chain's Sustainable Competitive Advantage. In R. Addo-Tenkorang, J. Kantola, P. Helo, & A. Shamsuzzoha (Eds.), *Supply Chain Strategies and the Engineer-to-Order Approach* (pp. 43–59). Hershey, PA: IGI Global. doi:10.4018/978-1-5225-0021-6.ch003

Adebiyi, I. D., Popoola, P. A., & Pityana, S. (2016). Mitigation of Wear Damage by Laser Surface Alloying Technique. In E. Akinlabi, R. Mahamood, & S. Akinlabi (Eds.), *Advanced Manufacturing Techniques Using Laser Material Processing* (pp. 172–196). Hershey, PA: IGI Global. doi:10.4018/978-1-5225-0329-3.ch007

Ahmad, W. (2016). Sulfur in Petroleum: Petroleum Desulfurization Techniques. In T. Saleh (Ed.), *Applying Nanotechnology to the Desulfurization Process in Petroleum Engineering* (pp. 1–52). Hershey, PA: IGI Global. doi:10.4018/978-1-4666-9545-0.ch001

Ahmed, I., Ahmad, N., Mehmood, I., Haq, I. U., Hassan, M., & Khan, M. U. (2016). Applications of Nanotechnology in Transportation Engineering. In A. Khitab & W. Anwar (Eds.), *Advanced Research on Nanotechnology for Civil Engineering Applications* (pp. 180–207). Hershey, PA: IGI Global. doi:10.4018/978-1-5225-0344-6.ch006

Aikhuele, D. (2018). A Study of Product Development Engineering and Design Reliability Concerns. *International Journal of Applied Industrial Engineering*, *5*(1), 79–89. doi:10.4018/IJAIE.2018010105

Al-Najar, B. T., & Bououdina, M. (2016). Bioinspired Nanoparticles for Efficient Drug Delivery System. In M. Bououdina (Ed.), *Emerging Research on Bioinspired Materials Engineering* (pp. 69–103). Hershey, PA: IGI Global. doi:10.4018/978-1-4666-9811-6.ch003

Al-Shebeeb, O. A., Rangaswamy, S., Gopalakrishan, B., & Devaru, D. G. (2017). Evaluation and Indexing of Process Plans Based on Electrical Demand and Energy Consumption. *International Journal of Manufacturing, Materials, and Mechanical Engineering*, *7*(3), 1–19. doi:10.4018/IJMMME.2017070101

Alexakis, H., & Makris, N. (2016). Validation of the Discrete Element Method for the Limit Stability Analysis of Masonry Arches. In V. Sarhosis, K. Bagi, J. Lemos, & G. Milani (Eds.), *Computational Modeling of Masonry Structures Using the Discrete Element Method* (pp. 292–325). Hershey, PA: IGI Global. doi:10.4018/978-1-5225-0231-9.ch012

AlMegren, H. A., Gonzalez-Cortes, S., Huang, Y., Chen, H., Qian, Y., Alkinany, M., ... Xiao, T. (2016). Preparation of Deep Hydrodesulfurzation Catalysts for Diesel Fuel using Organic Matrix Decomposition Method. In H. Al-Megren & T. Xiao (Eds.), *Petrochemical Catalyst Materials, Processes, and Emerging Technologies* (pp. 216–253). Hershey, PA: IGI Global. doi:10.4018/978-1-4666-9975-5.ch009

Alshammari, A., Kalevaru, V. N., Bagabas, A., & Martin, A. (2016). Production of Ethylene and its Commercial Importance in the Global Market. In H. Al-Megren & T. Xiao (Eds.), *Petrochemical Catalyst Materials, Processes, and Emerging Technologies* (pp. 82–115). Hershey, PA: IGI Global. doi:10.4018/978-1-4666-9975-5.ch004

Amel, M. (2016). Synthesis, Characterizations, and Biological Effects Study of Some Quinoline Family. In M. Bououdina (Ed.), *Emerging Research on Bioinspired Materials Engineering* (pp. 160–196). Hershey, PA: IGI Global. doi:10.4018/978-1-4666-9811-6.ch006

Amna, T., Haasan, M. S., Khil, M., & Hwang, I. (2016). Impact of Electrospun Biomimetic Extracellular Environment on Proliferation and Intercellular Communication of Muscle Precursor Cells: An Overview – Intercellular Communication of Muscle Precursor Cells with Extracellular Environment. In M. Bououdina (Ed.), *Emerging Research on Bioinspired Materials Engineering* (pp. 247–265). Hershey, PA: IGI Global. doi:10.4018/978-1-4666-9811-6.ch009

Amuda, M. O., Lawal, T. F., & Akinlabi, E. T. (2017). Research Progress on Rheological Behavior of AA7075 Aluminum Alloy During Hot Deformation. *International Journal of Materials Forming and Machining Processes*, *4*(1), 53–96. doi:10.4018/IJMFMP.2017010104

An, M., & Qin, Y. (2016). Challenges of Railway Safety Risk Assessment and Maintenance Decision Making. In B. Rai (Ed.), *Handbook of Research on Emerging Innovations in Rail Transportation Engineering* (pp. 173–211). Hershey, PA: IGI Global. doi:10.4018/978-1-5225-0084-1.ch009

Anil, M., Ayyildiz-Tamis, D., Tasdemir, S., Sendemir-Urkmez, A., & Gulce-Iz, S. (2016). Bioinspired Materials and Biocompatibility. In M. Bououdina (Ed.), *Emerging Research on Bioinspired Materials Engineering* (pp. 294–322). Hershey, PA: IGI Global. doi:10.4018/978-1-4666-9811-6.ch011

Armutlu, H. (2018). Intelligent Biomedical Engineering Operations by Cloud Computing Technologies. In U. Kose, G. Guraksin, & O. Deperlioglu (Eds.), *Nature-Inspired Intelligent Techniques for Solving Biomedical Engineering Problems* (pp. 297–317). Hershey, PA: IGI Global. doi:10.4018/978-1-5225-4769-3.ch015

Arokiyaraj, S., Saravanan, M., Bharanidharan, R., Islam, V. I., Bououdina, M., & Vincent, S. (2016). Green Synthesis of Metallic Nanoparticles Using Plant Compounds and Their Applications: Metallic Nanoparticles Synthesis Using Plants. In M. Bououdina (Ed.), *Emerging Research on Bioinspired Materials Engineering* (pp. 1–34). Hershey, PA: IGI Global. doi:10.4018/978-1-4666-9811-6.ch001

Atik, M., Sadek, M., & Shahrour, I. (2017). Single-Run Adaptive Pushover Procedure for Shear Wall Structures. In V. Plevris, G. Kremmyda, & Y. Fahjan (Eds.), *Performance-Based Seismic Design of Concrete Structures and Infrastructures* (pp. 59–83). Hershey, PA: IGI Global. doi:10.4018/978-1-5225-2089-4.ch003

Aydin, A., Akyol, E., Gungor, M., Kaya, A., & Tasdelen, S. (2018). Geophysical Surveys in Engineering Geology Investigations With Field Examples. In N. Ceryan (Ed.), *Handbook of Research on Trends and Digital Advances in Engineering Geology* (pp. 257–280). Hershey, PA: IGI Global. doi:10.4018/978-1-5225-2709-1.ch007

Azevedo, N. M., Lemos, J. V., & Rocha de Almeida, J. (2016). Discrete Element Particle Modelling of Stone Masonry. In V. Sarhosis, K. Bagi, J. Lemos, & G. Milani (Eds.), *Computational Modeling of Masonry Structures Using the Discrete Element Method* (pp. 146–170). Hershey, PA: IGI Global. doi:10.4018/978-1-5225-0231-9.ch007

Bamufleh, H. S., Noureldin, M. M., & El-Halwagi, M. M. (2016). Sustainable Process Integration in the Petrochemical Industries. In H. Al-Megren & T. Xiao (Eds.), *Petrochemical Catalyst Materials, Processes, and Emerging Technologies* (pp. 150–163). Hershey, PA: IGI Global. doi:10.4018/978-1-4666-9975-5.ch006

Banerjee, S., Gautam, R. K., Gautam, P. K., Jaiswal, A., & Chattopadhyaya, M. C. (2016). Recent Trends and Advancement in Nanotechnology for Water and Wastewater Treatment: Nanotechnological Approach for Water Purification. In A. Khitab & W. Anwar (Eds.), *Advanced Research on Nanotechnology for Civil Engineering Applications* (pp. 208–252). Hershey, PA: IGI Global. doi:10.4018/978-1-5225-0344-6.ch007

Bas, T. G. (2017). Nutraceutical Industry with the Collaboration of Biotechnology and Nutrigenomics Engineering: The Significance of Intellectual Property in the Entrepreneurship and Scientific Research Ecosystems. In T. Bas & J. Zhao (Eds.), *Comparative Approaches to Biotechnology Development and Use in Developed and Emerging Nations* (pp. 1–17). Hershey, PA: IGI Global. doi:10.4018/978-1-5225-1040-6.ch001

Beale, R., & André, J. (2017). *Design Solutions and Innovations in Temporary Structures*. Hershey, PA: IGI Global. doi:10.4018/978-1-5225-2199-0

Behnam, B. (2017). Simulating Post-Earthquake Fire Loading in Conventional RC Structures. In P. Samui, S. Chakraborty, & D. Kim (Eds.), *Modeling and Simulation Techniques in Structural Engineering* (pp. 425–444). Hershey, PA: IGI Global. doi:10.4018/978-1-5225-0588-4.ch015

Ben Hamida, I., Salah, S. B., Msahli, F., & Mimouni, M. F. (2018). Distribution Network Reconfiguration Using SPEA2 for Power Loss Minimization and Reliability Improvement. *International Journal of Energy Optimization and Engineering*, *7*(1), 50–65. doi:10.4018/IJEOE.2018010103

Benjamin, S. R., de Lima, F., & Rathoure, A. K. (2016). Genetically Engineered Microorganisms for Bioremediation Processes: GEMs for Bioremediaton. In A. Rathoure & V. Dhatwalia (Eds.), *Toxicity and Waste Management Using Bioremediation* (pp. 113–140). Hershey, PA: IGI Global. doi:10.4018/978-1-4666-9734-8.ch006

Bhaskar, S. V., & Kudal, H. N. (2017). Effect of TiCN and AlCrN Coating on Tribological Behaviour of Plasma-nitrided AISI 4140 Steel. *International Journal of Surface Engineering and Interdisciplinary Materials Science*, *5*(2), 1–17. doi:10.4018/IJSEIMS.2017070101

Bhowmik, S., Sahoo, P., Acharyya, S. K., Dhar, S., & Chattopadhyay, J. (2016). Effect of Microstructure Degradation on Fracture Toughness of 20MnMoNi55 Steel in DBT Region. *International Journal of Manufacturing, Materials, and Mechanical Engineering*, *6*(3), 11–27. doi:10.4018/IJMMME.2016070102

Bhutto, A. W., Abro, R., Abbas, T., Yu, G., & Chen, X. (2016). Desulphurization of Fuel Oils Using Ionic Liquids. In H. Al-Megren & T. Xiao (Eds.), *Petrochemical Catalyst Materials, Processes, and Emerging Technologies* (pp. 254–284). Hershey, PA: IGI Global. doi:10.4018/978-1-4666-9975-5.ch010

Bhuyan, D. (2018). Designing of a Twin Tube Shock Absorber: A Study in Reverse Engineering. In K. Kumar & J. Davim (Eds.), *Design and Optimization of Mechanical Engineering Products* (pp. 83–104). Hershey, PA: IGI Global. doi:10.4018/978-1-5225-3401-3.ch005

Bouloudenine, M., & Bououdina, M. (2016). Toxic Effects of Engineered Nanoparticles on Living Cells. In M. Bououdina (Ed.), *Emerging Research on Bioinspired Materials Engineering* (pp. 35–68). Hershey, PA: IGI Global. doi:10.4018/978-1-4666-9811-6.ch002

Brunetti, A., Sellaro, M., Drioli, E., & Barbieri, G. (2016). Membrane Engineering and its Role in Oil Refining and Petrochemical Industry. In H. Al-Megren & T. Xiao (Eds.), *Petrochemical Catalyst Materials, Processes, and Emerging Technologies* (pp. 116–149). Hershey, PA: IGI Global. doi:10.4018/978-1-4666-9975-5.ch005

Bügler, M., & Borrmann, A. (2016). Simulation Based Construction Project Schedule Optimization: An Overview on the State-of-the-Art. In F. Miranda & C. Abreu (Eds.), *Handbook of Research on Computational Simulation and Modeling in Engineering* (pp. 482–507). Hershey, PA: IGI Global. doi:10.4018/978-1-4666-8823-0.ch016

Calderon, F. A., Giolo, E. G., Frau, C. D., Rengel, M. G., Rodriguez, H., Tornello, M., ... Gallucci, R. (2018). Seismic Microzonation and Site Effects Detection Through Microtremors Measures: A Review. In N. Ceryan (Ed.), *Handbook of Research on Trends and Digital Advances in Engineering Geology* (pp. 326–349). Hershey, PA: IGI Global. doi:10.4018/978-1-5225-2709-1.ch009

Carmona-Murillo, J., & Valenzuela-Valdés, J. F. (2016). Motivation on Problem Based Learning. In D. Fonseca & E. Redondo (Eds.), *Handbook of Research on Applied E-Learning in Engineering and Architecture Education* (pp. 179–203). Hershey, PA: IGI Global. doi:10.4018/978-1-4666-8803-2.ch009

Ceryan, N. (2016). A Review of Soft Computing Methods Application in Rock Mechanic Engineering. In P. Samui (Ed.), *Handbook of Research on Advanced Computational Techniques for Simulation-Based Engineering* (pp. 1–70). Hershey, PA: IGI Global. doi:10.4018/978-1-4666-9479-8.ch001

Ceryan, N., & Can, N. K. (2018). Prediction of The Uniaxial Compressive Strength of Rocks Materials. In N. Ceryan (Ed.), *Handbook of Research on Trends and Digital Advances in Engineering Geology* (pp. 31–96). Hershey, PA: IGI Global. doi:10.4018/978-1-5225-2709-1.ch002

Ceryan, S. (2018). Weathering Indices Used in Evaluation of the Weathering State of Rock Material. In N. Ceryan (Ed.), *Handbook of Research on Trends and Digital Advances in Engineering Geology* (pp. 132–186). Hershey, PA: IGI Global. doi:10.4018/978-1-5225-2709-1.ch004

Chandrasekaran, S., Silva, B., Patil, A., Oo, A. M., & Campbell, M. (2016). Evaluating Engineering Students' Perceptions: The Impact of Team-Based Learning Practices in Engineering Education. *International Journal of Quality Assurance in Engineering and Technology Education, 5*(4), 42–59. doi:10.4018/IJQAETE.2016100103

Chen, H., Padilla, R. V., & Besarati, S. (2017). Supercritical Fluids and Their Applications in Power Generation. In L. Chen & Y. Iwamoto (Eds.), *Advanced Applications of Supercritical Fluids in Energy Systems* (pp. 369–402). Hershey, PA: IGI Global. doi:10.4018/978-1-5225-2047-4.ch012

Chen, L. (2017). Principles, Experiments, and Numerical Studies of Supercritical Fluid Natural Circulation System. In L. Chen & Y. Iwamoto (Eds.), *Advanced Applications of Supercritical Fluids in Energy Systems* (pp. 136–187). Hershey, PA: IGI Global. doi:10.4018/978-1-5225-2047-4.ch005

Clementi, F., Di Sciascio, G., Di Sciascio, S., & Lenci, S. (2017). Influence of the Shear-Bending Interaction on the Global Capacity of Reinforced Concrete Frames: A Brief Overview of the New Perspectives. In V. Plevris, G. Kremmyda, & Y. Fahjan (Eds.), *Performance-Based Seismic Design of Concrete Structures and Infrastructures* (pp. 84–111). Hershey, PA: IGI Global. doi:10.4018/978-1-5225-2089-4.ch004

Cortés-Polo, D., Calle-Cancho, J., Carmona-Murillo, J., & González-Sánchez, J. (2017). Future Trends in Mobile-Fixed Integration for Next Generation Networks: Classification and Analysis. *International Journal of Vehicular Telematics and Infotainment Systems, 1*(1), 33–53. doi:10.4018/IJVTIS.2017010103

Cui, X., Zeng, S., Li, Z., Zheng, Q., Yu, X., & Han, B. (2018). Advanced Composites for Civil Engineering Infrastructures. In K. Kumar & J. Davim (Eds.), *Composites and Advanced Materials for Industrial Applications* (pp. 212–248). Hershey, PA: IGI Global. doi:10.4018/978-1-5225-5216-1.ch010

Dalgıç, S., & Kuşku, İ. (2018). Geological and Geotechnical Investigations in Tunneling. In N. Ceryan (Ed.), *Handbook of Research on Trends and Digital Advances in Engineering Geology* (pp. 482–529). Hershey, PA: IGI Global. doi:10.4018/978-1-5225-2709-1.ch014

de la Varga, D., Soto, M., Arias, C. A., van Oirschot, D., Kilian, R., Pascual, A., & Álvarez, J. A. (2017). Constructed Wetlands for Industrial Wastewater Treatment and Removal of Nutrients. In Á. Val del Río, J. Campos Gómez, & A. Mosquera Corral (Eds.), *Technologies for the Treatment and Recovery of Nutrients from Industrial Wastewater* (pp. 202–230). Hershey, PA: IGI Global. doi:10.4018/978-1-5225-1037-6.ch008

del Valle-Zermeño, R., Chimenos, J. M., & Formosa, J. (2016). Flue Gas Desulfurization: Processes and Technologies. In T. Saleh (Ed.), *Applying Nanotechnology to the Desulfurization Process in Petroleum Engineering* (pp. 337–377). Hershey, PA: IGI Global. doi:10.4018/978-1-4666-9545-0.ch011

Delgado, J. M., Henriques, A. A., & Delgado, R. M. (2016). Structural Non-Linear Models and Simulation Techniques: An Efficient Combination for Safety Evaluation of RC Structures. In F. Miranda & C. Abreu (Eds.), *Handbook of Research on Computational Simulation and Modeling in Engineering* (pp. 540–584). Hershey, PA: IGI Global. doi:10.4018/978-1-4666-8823-0.ch018

Delgado, P. S., Arêde, A., Pouca, N. V., & Costa, A. (2016). Numerical Modeling of RC Bridges for Seismic Risk Analysis. In F. Miranda & C. Abreu (Eds.), *Handbook of Research on Computational Simulation and Modeling in Engineering* (pp. 457–481). Hershey, PA: IGI Global. doi:10.4018/978-1-4666-8823-0.ch015

Deng, Y., & Liu, S. (2016). Catalysis with Room Temperature Ionic Liquids Mediated Metal Nanoparticles. In H. Al-Megren & T. Xiao (Eds.), *Petrochemical Catalyst Materials, Processes, and Emerging Technologies* (pp. 285–329). Hershey, PA: IGI Global. doi:10.4018/978-1-4666-9975-5.ch011

Deperlioglu, O. (2018). Intelligent Techniques Inspired by Nature and Used in Biomedical Engineering. In U. Kose, G. Guraksin, & O. Deperlioglu (Eds.), *Nature-Inspired Intelligent Techniques for Solving Biomedical Engineering Problems* (pp. 51–77). Hershey, PA: IGI Global. doi:10.4018/978-1-5225-4769-3.ch003

Dias, G. L., Magalhães, R. R., Ferreira, D. D., & Vitoriano, F. A. (2016). The Use of a Robotic Arm for Displacement Measurements in a Cantilever beam. *International Journal of Manufacturing, Materials, and Mechanical Engineering*, *6*(3), 45–57. doi:10.4018/IJMMME.2016070104

Dimitratos, N., Villa, A., Chan-Thaw, C. E., Hammond, C., & Prati, L. (2016). Valorisation of Glycerol to Fine Chemicals and Fuels. In H. Al-Megren & T. Xiao (Eds.), *Petrochemical Catalyst Materials, Processes, and Emerging Technologies* (pp. 352–384). Hershey, PA: IGI Global. doi:10.4018/978-1-4666-9975-5.ch013

Dixit, A. (2018). Application of Silica-Gel-Reinforced Aluminium Composite on the Piston of Internal Combustion Engine: Comparative Study of Silica-Gel-Reinforced Aluminium Composite Piston With Aluminium Alloy Piston. In K. Kumar & J. Davim (Eds.), *Composites and Advanced Materials for Industrial Applications* (pp. 63–98). Hershey, PA: IGI Global. doi:10.4018/978-1-5225-5216-1.ch004

Drei, A., Milani, G., & Sincraian, G. (2016). Application of DEM to Historic Masonries, Two Case-Studies in Portugal and Italy: Aguas Livres Aqueduct and Arch-Tympana of a Church. In V. Sarhosis, K. Bagi, J. Lemos, & G. Milani (Eds.), *Computational Modeling of Masonry Structures Using the Discrete Element Method* (pp. 326–366). Hershey, PA: IGI Global. doi:10.4018/978-1-5225-0231-9.ch013

Dutta, S., Roy, P. K., & Nandi, D. (2016). Optimal Allocation of Static Synchronous Series Compensator Controllers using Chemical Reaction Optimization for Reactive Power Dispatch. *International Journal of Energy Optimization and Engineering*, *5*(3), 43–62. doi:10.4018/IJEOE.2016070103

Dutta, S., Roy, P. K., & Nandi, D. (2016). Quasi Oppositional Teaching-Learning based Optimization for Optimal Power Flow Incorporating FACTS. *International Journal of Energy Optimization and Engineering*, *5*(2), 64–84. doi:10.4018/IJEOE.2016040104

Eloy, S., Dias, M. S., Lopes, P. F., & Vilar, E. (2016). Digital Technologies in Architecture and Engineering: Exploring an Engaged Interaction within Curricula. In D. Fonseca & E. Redondo (Eds.), *Handbook of Research on Applied E-Learning in Engineering and Architecture Education* (pp. 368–402). Hershey, PA: IGI Global. doi:10.4018/978-1-4666-8803-2.ch017

Elsayed, A. M., Dakkama, H. J., Mahmoud, S., Al-Dadah, R., & Kaialy, W. (2017). Sustainable Cooling Research Using Activated Carbon Adsorbents and Their Environmental Impact. In T. Kobayashi (Ed.), *Applied Environmental Materials Science for Sustainability* (pp. 186–221). Hershey, PA: IGI Global. doi:10.4018/978-1-5225-1971-3.ch009

Ercanoglu, M., & Sonmez, H. (2018). General Trends and New Perspectives on Landslide Mapping and Assessment Methods. In N. Ceryan (Ed.), *Handbook of Research on Trends and Digital Advances in Engineering Geology* (pp. 350–379). Hershey, PA: IGI Global. doi:10.4018/978-1-5225-2709-1.ch010

Erinosho, M. F., Akinlabi, E. T., & Pityana, S. (2016). Enhancement of Surface Integrity of Titanium Alloy with Copper by Means of Laser Metal Deposition Process. In E. Akinlabi, R. Mahamood, & S. Akinlabi (Eds.), *Advanced Manufacturing Techniques Using Laser Material Processing* (pp. 60–91). Hershey, PA: IGI Global. doi:10.4018/978-1-5225-0329-3.ch004

Farag, H., & Kishida, M. (2016). Kinetic Models for Complex Parallel–Consecutive Reactions Assessment of Reaction Network and Product Selectivity. In H. Al-Megren & T. Xiao (Eds.), *Petrochemical Catalyst Materials, Processes, and Emerging Technologies* (pp. 330–351). Hershey, PA: IGI Global. doi:10.4018/978-1-4666-9975-5.ch012

Faroz, S. A., Pujari, N. N., Rastogi, R., & Ghosh, S. (2017). Risk Analysis of Structural Engineering Systems Using Bayesian Inference. In P. Samui, S. Chakraborty, & D. Kim (Eds.), *Modeling and Simulation Techniques in Structural Engineering* (pp. 390–424). Hershey, PA: IGI Global. doi:10.4018/978-1-5225-0588-4.ch014

Fernando, P. R., Hamigah, T., Disne, S., Wickramasingha, G. G., & Sutharshan, A. (2018). The Evaluation of Engineering Properties of Low Cost Concrete Blocks by Partial Doping of Sand with Sawdust: Low Cost Sawdust Concrete Block. *International Journal of Strategic Engineering*, *1*(2), 26–42. doi:10.4018/IJoSE.2018070103

Fragiadakis, M., Stefanou, I., & Psycharis, I. N. (2016). Vulnerability Assessment of Damaged Classical Multidrum Columns. In V. Sarhosis, K. Bagi, J. Lemos, & G. Milani (Eds.), *Computational Modeling of Masonry Structures Using the Discrete Element Method* (pp. 235–253). Hershey, PA: IGI Global. doi:10.4018/978-1-5225-0231-9.ch010

Gaines, T. W., Williams, K. R., & Wagener, K. B. (2016). ADMET: Functionalized Polyolefins. In H. Al-Megren & T. Xiao (Eds.), *Petrochemical Catalyst Materials, Processes, and Emerging Technologies* (pp. 1–21). Hershey, PA: IGI Global. doi:10.4018/978-1-4666-9975-5.ch001

Garg, H. (2016). Bi-Criteria Optimization for Finding the Optimal Replacement Interval for Maintaining the Performance of the Process Industries. In P. Vasant, G. Weber, & V. Dieu (Eds.), *Handbook of Research on Modern Optimization Algorithms and Applications in Engineering and Economics* (pp. 643–675). Hershey, PA: IGI Global. doi:10.4018/978-1-4666-9644-0.ch025

Gaspar, P. D., Dinho da Silva, P., Gonçalves, J. P., & Carneiro, R. (2016). Computational Modelling and Simulation to Assist the Improvement of Thermal Performance and Energy Efficiency in Industrial Engineering Systems: Application to Cold Stores. In F. Miranda & C. Abreu (Eds.), *Handbook of Research on Computational Simulation and Modeling in Engineering* (pp. 1–68). Hershey, PA: IGI Global. doi:10.4018/978-1-4666-8823-0.ch001

Ge, H., Tang, M., & Wen, X. (2016). Ni/ZnO Nano Sorbent for Reactive Adsorption Desulfurization of Refinery Oil Streams. In T. Saleh (Ed.), *Applying Nanotechnology to the Desulfurization Process in Petroleum Engineering* (pp. 216–239). Hershey, PA: IGI Global. doi:10.4018/978-1-4666-9545-0.ch007

Ghosh, S., Mitra, S., Ghosh, S., & Chakraborty, S. (2017). Seismic Reliability Analysis in the Framework of Metamodelling Based Monte Carlo Simulation. In P. Samui, S. Chakraborty, & D. Kim (Eds.), *Modeling and Simulation Techniques in Structural Engineering* (pp. 192–208). Hershey, PA: IGI Global. doi:10.4018/978-1-5225-0588-4.ch006

Gil, M., & Otero, B. (2017). Learning Engineering Skills through Creativity and Collaboration: A Game-Based Proposal. In R. Alexandre Peixoto de Queirós & M. Pinto (Eds.), *Gamification-Based E-Learning Strategies for Computer Programming Education* (pp. 14–29). Hershey, PA: IGI Global. doi:10.4018/978-1-5225-1034-5.ch002

Gill, J., Ayre, M., & Mills, J. (2017). Revisioning the Engineering Profession: How to Make It Happen! In M. Gray & K. Thomas (Eds.), *Strategies for Increasing Diversity in Engineering Majors and Careers* (pp. 156–175). Hershey, PA: IGI Global. doi:10.4018/978-1-5225-2212-6.ch008

Gopal, S., & Al-Hazmi, M. H. (2016). Advances in Catalytic Technologies for Selective Oxidation of Lower Alkanes. In H. Al-Megren & T. Xiao (Eds.), *Petrochemical Catalyst Materials, Processes, and Emerging Technologies* (pp. 22–52). Hershey, PA: IGI Global. doi:10.4018/978-1-4666-9975-5.ch002

Goyal, N., Ram, M., Bhardwaj, A., & Kumar, A. (2016). Thermal Power Plant Modelling with Fault Coverage Stochastically. *International Journal of Manufacturing, Materials, and Mechanical Engineering*, *6*(3), 28–44. doi:10.4018/IJMMME.2016070103

Goyal, N., Ram, M., & Kumar, P. (2017). Welding Process under Fault Coverage Approach for Reliability and MTTF. In M. Ram & J. Davim (Eds.), *Mathematical Concepts and Applications in Mechanical Engineering and Mechatronics* (pp. 222–245). Hershey, PA: IGI Global. doi:10.4018/978-1-5225-1639-2.ch011

Gray, M., & Lundy, C. (2017). Engineering Study Abroad: High Impact Strategy for Increasing Access. In M. Gray & K. Thomas (Eds.), *Strategies for Increasing Diversity in Engineering Majors and Careers* (pp. 42–59). Hershey, PA: IGI Global. doi:10.4018/978-1-5225-2212-6.ch003

Guha, D., Roy, P. K., & Banerjee, S. (2016). Application of Modified Biogeography Based Optimization in AGC of an Interconnected Multi-Unit Multi-Source AC-DC Linked Power System. *International Journal of Energy Optimization and Engineering*, *5*(3), 1–18. doi:10.4018/IJEOE.2016070101

Guha, D., Roy, P. K., & Banerjee, S. (2016). Grey Wolf Optimization to Solve Load Frequency Control of an Interconnected Power System: GWO Used to Solve LFC Problem. *International Journal of Energy Optimization and Engineering*, *5*(4), 62–83. doi:10.4018/IJEOE.2016100104

Gupta, A. K., Dey, A., & Mukhopadhyay, A. K. (2016). Micromechanical and Finite Element Modeling for Composites. In S. Datta & J. Davim (Eds.), *Computational Approaches to Materials Design: Theoretical and Practical Aspects* (pp. 101–162). Hershey, PA: IGI Global. doi:10.4018/978-1-5225-0290-6.ch005

Guraksin, G. E. (2018). Internet of Things and Nature-Inspired Intelligent Techniques for the Future of Biomedical Engineering. In U. Kose, G. Guraksin, & O. Deperlioglu (Eds.), *Nature-Inspired Intelligent Techniques for Solving Biomedical Engineering Problems* (pp. 263–282). Hershey, PA: IGI Global. doi:10.4018/978-1-5225-4769-3.ch013

Hansman, C. A. (2016). Developing Mentoring Programs in Engineering and Technology Education. *International Journal of Quality Assurance in Engineering and Technology Education*, *5*(2), 1–15. doi:10.4018/IJQAETE.2016040101

Hasan, U., Chegenizadeh, A., & Nikraz, H. (2016). Nanotechnology Future and Present in Construction Industry: Applications in Geotechnical Engineering. In A. Khitab & W. Anwar (Eds.), *Advanced Research on Nanotechnology for Civil Engineering Applications* (pp. 141–179). Hershey, PA: IGI Global. doi:10.4018/978-1-5225-0344-6.ch005

Hejazi, T., & Akbari, L. (2017). A Multiresponse Optimization Model for Statistical Design of Processes with Discrete Variables. In M. Ram & J. Davim (Eds.), *Mathematical Concepts and Applications in Mechanical Engineering and Mechatronics* (pp. 17–37). Hershey, PA: IGI Global. doi:10.4018/978-1-5225-1639-2.ch002

Hejazi, T., & Hejazi, A. (2017). Monte Carlo Simulation for Reliability-Based Design of Automotive Complex Subsystems. In M. Ram & J. Davim (Eds.), *Mathematical Concepts and Applications in Mechanical Engineering and Mechatronics* (pp. 177–200). Hershey, PA: IGI Global. doi:10.4018/978-1-5225-1639-2.ch009

Hejazi, T., & Poursabbagh, H. (2017). Reliability Analysis of Engineering Systems: An Accelerated Life Testing for Boiler Tubes. In M. Ram & J. Davim (Eds.), *Mathematical Concepts and Applications in Mechanical Engineering and Mechatronics* (pp. 154–176). Hershey, PA: IGI Global. doi:10.4018/978-1-5225-1639-2.ch008

Henao, J., & Sotelo, O. (2018). Surface Engineering at High Temperature: Thermal Cycling and Corrosion Resistance. In A. Pakseresht (Ed.), *Production, Properties, and Applications of High Temperature Coatings* (pp. 131–159). Hershey, PA: IGI Global. doi:10.4018/978-1-5225-4194-3.ch006

Huirache-Acuña, R., Alonso-Nuñez, G., Rivera-Muñoz, E. M., Gutierrez, O., & Pawelec, B. (2016). Trimetallic Sulfide Catalysts for Hydrodesulfurization. In T. Saleh (Ed.), *Applying Nanotechnology to the Desulfurization Process in Petroleum Engineering* (pp. 240–262). Hershey, PA: IGI Global. doi:10.4018/978-1-4666-9545-0.ch008

Ilori, O. O., Adetan, D. A., & Umoru, L. E. (2017). Effect of Cutting Parameters on the Surface Residual Stress of Face-Milled Pearlitic Ductile Iron. *International Journal of Materials Forming and Machining Processes*, *4*(1), 38–52. doi:10.4018/IJMFMP.2017010103

Imam, M. H., Tasadduq, I. A., Ahmad, A., Aldosari, F., & Khan, H. (2017). Automated Generation of Course Improvement Plans Using Expert System. *International Journal of Quality Assurance in Engineering and Technology Education*, *6*(1), 1–12. doi:10.4018/IJQAETE.2017010101

Injeti, S. K., & Kumar, T. V. (2018). A WDO Framework for Optimal Deployment of DGs and DSCs in a Radial Distribution System Under Daily Load Pattern to Improve Techno-Economic Benefits. *International Journal of Energy Optimization and Engineering*, *7*(2), 1–38. doi:10.4018/IJEOE.2018040101

Ishii, N., Anami, K., & Knisely, C. W. (2018). *Dynamic Stability of Hydraulic Gates and Engineering for Flood Prevention*. Hershey, PA: IGI Global. doi:10.4018/978-1-5225-3079-4

J., J., Chowdhury, S., Goyal, P., Samui, P., & Dalkiliç, Y. (2016). Determination of Bearing Capacity of Shallow Foundation Using Soft Computing. In P. Saxena, D. Singh, & M. Pant (Eds.), *Problem Solving and Uncertainty Modeling through Optimization and Soft Computing Applications* (pp. 292-328). Hershey, PA: IGI Global. doi:10.4018/978-1-4666-9885-7.ch014

Jagan, J., Gundlapalli, P., & Samui, P. (2016). Utilization of Classification Techniques for the Determination of Liquefaction Susceptibility of Soils. In S. Bhattacharyya, P. Banerjee, D. Majumdar, & P. Dutta (Eds.), *Handbook of Research on Advanced Hybrid Intelligent Techniques and Applications* (pp. 124–160). Hershey, PA: IGI Global. doi:10.4018/978-1-4666-9474-3.ch005

Jayapalan, S. (2018). A Review of Chemical Treatments on Natural Fibers-Based Hybrid Composites for Engineering Applications. In K. Kumar & J. Davim (Eds.), *Composites and Advanced Materials for Industrial Applications* (pp. 16–37). Hershey, PA: IGI Global. doi:10.4018/978-1-5225-5216-1.ch002

Jeet, K., & Dhir, R. (2016). Software Module Clustering Using Bio-Inspired Algorithms. In P. Vasant, G. Weber, & V. Dieu (Eds.), *Handbook of Research on Modern Optimization Algorithms and Applications in Engineering and Economics* (pp. 445–470). Hershey, PA: IGI Global. doi:10.4018/978-1-4666-9644-0.ch017

Joshi, S. D., & Talange, D. B. (2016). Fault Tolerant Control for a Fractional Order AUV System. *International Journal of Energy Optimization and Engineering*, *5*(2), 1–24. doi:10.4018/IJEOE.2016040101

Julião, D., Ribeiro, S., de Castro, B., Cunha-Silva, L., & Balula, S. S. (2016). Polyoxometalates-Based Nanocatalysts for Production of Sulfur-Free Diesel. In T. Saleh (Ed.), *Applying Nanotechnology to the Desulfurization Process in Petroleum Engineering* (pp. 426–458). Hershey, PA: IGI Global. doi:10.4018/978-1-4666-9545-0.ch014

Kamthan, P. (2016). On the Nature of Collaborations in Agile Software Engineering Course Projects. *International Journal of Quality Assurance in Engineering and Technology Education*, *5*(2), 42–59. doi:10.4018/IJQAETE.2016040104

Karaman, O., Celik, C., & Urkmez, A. S. (2016). Self-Assembled Biomimetic Scaffolds for Bone Tissue Engineering. In M. Bououdina (Ed.), *Emerging Research on Bioinspired Materials Engineering* (pp. 104–132). Hershey, PA: IGI Global. doi:10.4018/978-1-4666-9811-6.ch004

Karkalos, N. E., Markopoulos, A. P., & Dossis, M. F. (2017). Optimal Model Parameters of Inverse Kinematics Solution of a 3R Robotic Manipulator Using ANN Models. *International Journal of Manufacturing, Materials, and Mechanical Engineering*, *7*(3), 20–40. doi:10.4018/IJMMME.2017070102

Kesimal, A., Karaman, K., Cihangir, F., & Ercikdi, B. (2018). Excavatability Assessment of Rock Masses for Geotechnical Studies. In N. Ceryan (Ed.), *Handbook of Research on Trends and Digital Advances in Engineering Geology* (pp. 231–256). Hershey, PA: IGI Global. doi:10.4018/978-1-5225-2709-1.ch006

Khanh, D. V., Vasant, P. M., Elamvazuthi, I., & Dieu, V. N. (2016). Multi-Objective Optimization of Two-Stage Thermo-Electric Cooler Using Differential Evolution: MO Optimization of TEC Using DE. In F. Miranda & C. Abreu (Eds.), *Handbook of Research on Computational Simulation and Modeling in Engineering* (pp. 139–170). Hershey, PA: IGI Global. doi:10.4018/978-1-4666-8823-0.ch004

Kim, D., Hassan, M. K., Chang, S., & Bigdeli, Y. (2016). Nonlinear Vibration Control of 3D Irregular Structures Subjected to Seismic Loads. In P. Samui (Ed.), *Handbook of Research on Advanced Computational Techniques for Simulation-Based Engineering* (pp. 103–119). Hershey, PA: IGI Global. doi:10.4018/978-1-4666-9479-8.ch003

Knoflacher, H. (2017). The Role of Engineers and Their Tools in the Transport Sector after Paradigm Change: From Assumptions and Extrapolations to Science. In H. Knoflacher & E. Ocalir-Akunal (Eds.), *Engineering Tools and Solutions for Sustainable Transportation Planning* (pp. 1–29). Hershey, PA: IGI Global. doi:10.4018/978-1-5225-2116-7.ch001

Kose, U. (2018). Towards an Intelligent Biomedical Engineering With Nature-Inspired Artificial Intelligence Techniques. In U. Kose, G. Guraksin, & O. Deperlioglu (Eds.), *Nature-Inspired Intelligent Techniques for Solving Biomedical Engineering Problems* (pp. 1–26). Hershey, PA: IGI Global. doi:10.4018/978-1-5225-4769-3.ch001

Kostić, S. (2018). A Review on Enhanced Stability Analyses of Soil Slopes Using Statistical Design. In N. Ceryan (Ed.), *Handbook of Research on Trends and Digital Advances in Engineering Geology* (pp. 446–481). Hershey, PA: IGI Global. doi:10.4018/978-1-5225-2709-1.ch013

Kumar, A., Patil, P. P., & Prajapati, Y. K. (2018). *Advanced Numerical Simulations in Mechanical Engineering*. Hershey, PA: IGI Global. doi:10.4018/978-1-5225-3722-9

Kumar, G. R., Rajyalakshmi, G., & Manupati, V. K. (2017). Surface Micro Patterning of Aluminium Reinforced Composite through Laser Peening. *International Journal of Manufacturing, Materials, and Mechanical Engineering*, *7*(4), 15–27. doi:10.4018/IJMMME.2017100102

Kumari, N., & Kumar, K. (2018). Fabrication of Orthotic Calipers With Epoxy-Based Green Composite. In K. Kumar & J. Davim (Eds.), *Composites and Advanced Materials for Industrial Applications* (pp. 157–176). Hershey, PA: IGI Global. doi:10.4018/978-1-5225-5216-1.ch008

Kuppusamy, R. R. (2018). Development of Aerospace Composite Structures Through Vacuum-Enhanced Resin Transfer Moulding Technology (VERTMTy): Vacuum-Enhanced Resin Transfer Moulding. In K. Kumar & J. Davim (Eds.), *Composites and Advanced Materials for Industrial Applications* (pp. 99–111). Hershey, PA: IGI Global. doi:10.4018/978-1-5225-5216-1.ch005

Lemos, J. V. (2016). The Basis for Masonry Analysis with UDEC and 3DEC. In V. Sarhosis, K. Bagi, J. Lemos, & G. Milani (Eds.), *Computational Modeling of Masonry Structures Using the Discrete Element Method* (pp. 61–89). Hershey, PA: IGI Global. doi:10.4018/978-1-5225-0231-9.ch003

Loy, J., Howell, S., & Cooper, R. (2017). Engineering Teams: Supporting Diversity in Engineering Education. In M. Gray & K. Thomas (Eds.), *Strategies for Increasing Diversity in Engineering Majors and Careers* (pp. 106–129). Hershey, PA: IGI Global. doi:10.4018/978-1-5225-2212-6.ch006

Macher, G., Armengaud, E., Kreiner, C., Brenner, E., Schmittner, C., Ma, Z., ... Krammer, M. (2018). Integration of Security in the Development Lifecycle of Dependable Automotive CPS. In N. Druml, A. Genser, A. Krieg, M. Menghin, & A. Hoeller (Eds.), *Solutions for Cyber-Physical Systems Ubiquity* (pp. 383–423). Hershey, PA: IGI Global. doi:10.4018/978-1-5225-2845-6.ch015

Maghsoodlou, S., & Poreskandar, S. (2016). Controlling Electrospinning Jet Using Microscopic Model for Ideal Tissue Engineering Scaffolds. *International Journal of Chemoinformatics and Chemical Engineering*, *5*(2), 1–16. doi:10.4018/IJCCE.2016070101

Mahendramani, G., & Lakshmana Swamy, N. (2018). Effect of Weld Groove Area on Distortion of Butt Welded Joints in Submerged Arc Welding. *International Journal of Manufacturing, Materials, and Mechanical Engineering*, *8*(2), 33–44. doi:10.4018/IJMMME.2018040103

Maiti, S. (2016). Engineered Gellan Polysaccharides in the Design of Controlled Drug Delivery Systems. In M. Bououdina (Ed.), *Emerging Research on Bioinspired Materials Engineering* (pp. 266–293). Hershey, PA: IGI Global. doi:10.4018/978-1-4666-9811-6.ch010

Majumdar, J. D., Weisheit, A., & Manna, I. (2016). Laser Surface Processing for Tailoring of Properties by Optimization of Microstructure. In E. Akinlabi, R. Mahamood, & S. Akinlabi (Eds.), *Advanced Manufacturing Techniques Using Laser Material Processing* (pp. 121–171). Hershey, PA: IGI Global. doi:10.4018/978-1-5225-0329-3.ch006

Maldonado-Macías, A. A., García-Alcaraz, J. L., Hernández-Arellano, J. L., & Cortes-Robles, G. (2016). An Ergonomic Compatibility Perspective on the Selection of Advanced Manufacturing Technology: A Case Study for CNC Vertical Machining Centers. In G. Alor-Hernández, C. Sánchez-Ramírez, & J. García-Alcaraz (Eds.), *Handbook of Research on Managerial Strategies for Achieving Optimal Performance in Industrial Processes* (pp. 137–165). Hershey, PA: IGI Global. doi:10.4018/978-1-5225-0130-5.ch008

Mamaghani, I. H. (2016). Application of Discrete Finite Element Method for Analysis of Unreinforced Masonry Structures. In V. Sarhosis, K. Bagi, J. Lemos, & G. Milani (Eds.), *Computational Modeling of Masonry Structures Using the Discrete Element Method* (pp. 440–458). Hershey, PA: IGI Global. doi:10.4018/978-1-5225-0231-9.ch017

Mansor, M. R., Sapuan, S. M., Salim, M. A., Akop, M. Z., Musthafah, M. T., & Shaharuzaman, M. A. (2016). Concurrent Design of Green Composites. In D. Verma, S. Jain, X. Zhang, & P. Gope (Eds.), *Green Approaches to Biocomposite Materials Science and Engineering* (pp. 48–75). Hershey, PA: IGI Global. doi:10.4018/978-1-5225-0424-5.ch003

Mansouri, I., & Esmaeili, E. (2016). Nanotechnology Applications in the Construction Industry. In A. Khitab & W. Anwar (Eds.), *Advanced Research on Nanotechnology for Civil Engineering Applications* (pp. 111–140). Hershey, PA: IGI Global. doi:10.4018/978-1-5225-0344-6.ch004

Manzoor, A. (2016). MOOCs for Enhancing Engineering Education. In D. Fonseca & E. Redondo (Eds.), *Handbook of Research on Applied E-Learning in Engineering and Architecture Education* (pp. 204–223). Hershey, PA: IGI Global. doi:10.4018/978-1-4666-8803-2.ch010

Martin, A., Kalevaru, V. N., & Radnik, J. (2016). Palladium in Heterogeneous Oxidation Catalysis. In H. Al-Megren & T. Xiao (Eds.), *Petrochemical Catalyst Materials, Processes, and Emerging Technologies* (pp. 53–81). Hershey, PA: IGI Global. doi:10.4018/978-1-4666-9975-5.ch003

Melnyczuk, J. M., & Palchoudhury, S. (2016). Introduction to Bio-Inspired Hydrogel and Their Application: Hydrogels. In M. Bououdina (Ed.), *Emerging Research on Bioinspired Materials Engineering* (pp. 133–159). Hershey, PA: IGI Global. doi:10.4018/978-1-4666-9811-6.ch005

Mitra-Kirtley, S., Mullins, O. C., & Pomerantz, A. E. (2016). Sulfur and Nitrogen Chemical Speciation in Crude Oils and Related Carbonaceous Materials. In T. Saleh (Ed.), *Applying Nanotechnology to the Desulfurization Process in Petroleum Engineering* (pp. 53–83). Hershey, PA: IGI Global. doi:10.4018/978-1-4666-9545-0.ch002

Moalosi, R., Uziak, J., & Oladiran, M. T. (2016). Using Blended Learning Approach to Deliver Courses in An Engineering Programme. *International Journal of Quality Assurance in Engineering and Technology Education*, *5*(1), 23–39. doi:10.4018/IJQAETE.2016010103

Mohammadzadeh, S., & Kim, Y. (2017). Nonlinear System Identification of Smart Buildings. In P. Samui, S. Chakraborty, & D. Kim (Eds.), *Modeling and Simulation Techniques in Structural Engineering* (pp. 328–347). Hershey, PA: IGI Global. doi:10.4018/978-1-5225-0588-4.ch011

Mohanty, I., & Bhattacherjee, D. (2016). Artificial Neural Network and Its Application in Steel Industry. In S. Datta & J. Davim (Eds.), *Computational Approaches to Materials Design: Theoretical and Practical Aspects* (pp. 267–300). Hershey, PA: IGI Global. doi:10.4018/978-1-5225-0290-6.ch010

Mohebkhah, A., & Sarhosis, V. (2016). Discrete Element Modeling of Masonry-Infilled Frames. In V. Sarhosis, K. Bagi, J. Lemos, & G. Milani (Eds.), *Computational Modeling of Masonry Structures Using the Discrete Element Method* (pp. 200–234). Hershey, PA: IGI Global. doi:10.4018/978-1-5225-0231-9.ch009

Molina, G. J., Aktaruzzaman, F., Soloiu, V., & Rahman, M. (2017). Design and Testing of a Jet-Impingement Instrument to Study Surface-Modification Effects by Nanofluids. *International Journal of Surface Engineering and Interdisciplinary Materials Science*, *5*(2), 43–61. doi:10.4018/IJSEIMS.2017070104

Montalvan-Sorrosa, D., de los Cobos-Vasconcelos, D., & Gonzalez-Sanchez, A. (2016). Nanotechnology Applied to the Biodesulfurization of Fossil Fuels and Spent Caustic Streams. In T. Saleh (Ed.), *Applying Nanotechnology to the Desulfurization Process in Petroleum Engineering* (pp. 378–389). Hershey, PA: IGI Global. doi:10.4018/978-1-4666-9545-0.ch012

Montillet, J., Yu, K., Bonenberg, L. K., & Roberts, G. W. (2016). Optimization Algorithms in Local and Global Positioning. In P. Vasant, G. Weber, & V. Dieu (Eds.), *Handbook of Research on Modern Optimization Algorithms and Applications in Engineering and Economics* (pp. 1–53). Hershey, PA: IGI Global. doi:10.4018/978-1-4666-9644-0.ch001

Moreira, F., & Ferreira, M. J. (2016). Teaching and Learning Requirements Engineering Based on Mobile Devices and Cloud: A Case Study. In D. Fonseca & E. Redondo (Eds.), *Handbook of Research on Applied E-Learning in Engineering and Architecture Education* (pp. 237–262). Hershey, PA: IGI Global. doi:10.4018/978-1-4666-8803-2.ch012

Mukherjee, A., Saeed, R. A., Dutta, S., & Naskar, M. K. (2017). Fault Tracking Framework for Software-Defined Networking (SDN). In C. Singhal & S. De (Eds.), *Resource Allocation in Next-Generation Broadband Wireless Access Networks* (pp. 247–272). Hershey, PA: IGI Global. doi:10.4018/978-1-5225-2023-8.ch011

Mukhopadhyay, A., Barman, T. K., & Sahoo, P. (2018). Electroless Nickel Coatings for High Temperature Applications. In K. Kumar & J. Davim (Eds.), *Composites and Advanced Materials for Industrial Applications* (pp. 297–331). Hershey, PA: IGI Global. doi:10.4018/978-1-5225-5216-1.ch013

Náprstek, J., & Fischer, C. (2017). Dynamic Stability and Post-Critical Processes of Slender Auto-Parametric Systems. In V. Plevris, G. Kremmyda, & Y. Fahjan (Eds.), *Performance-Based Seismic Design of Concrete Structures and Infrastructures* (pp. 128–171). Hershey, PA: IGI Global. doi:10.4018/978-1-5225-2089-4.ch006

Nautiyal, L., Shivach, P., & Ram, M. (2018). Optimal Designs by Means of Genetic Algorithms. In M. Ram & J. Davim (Eds.), *Soft Computing Techniques and Applications in Mechanical Engineering* (pp. 151–161). Hershey, PA: IGI Global. doi:10.4018/978-1-5225-3035-0.ch007

Nazir, R. (2017). Advanced Nanomaterials for Water Engineering and Treatment: Nano-Metal Oxides and Their Nanocomposites. In T. Saleh (Ed.), *Advanced Nanomaterials for Water Engineering, Treatment, and Hydraulics* (pp. 84–126). Hershey, PA: IGI Global. doi:10.4018/978-1-5225-2136-5.ch005

Nogueira, A. F., Ribeiro, J. C., Fernández de Vega, F., & Zenha-Rela, M. A. (2018). Evolutionary Approaches to Test Data Generation for Object-Oriented Software: Overview of Techniques and Tools. In M. Khosrow-Pour, D.B.A. (Ed.), Incorporating Nature-Inspired Paradigms in Computational Applications (pp. 162-194). Hershey, PA: IGI Global. doi:10.4018/978-1-5225-5020-4.ch006

Nunes, J. F., Moreira, P. M., & Tavares, J. M. (2016). Human Motion Analysis and Simulation Tools: A Survey. In F. Miranda & C. Abreu (Eds.), *Handbook of Research on Computational Simulation and Modeling in Engineering* (pp. 359–388). Hershey, PA: IGI Global. doi:10.4018/978-1-4666-8823-0.ch012

Ogunlaja, A. S., & Tshentu, Z. R. (2016). Molecularly Imprinted Polymer Nanofibers for Adsorptive Desulfurization. In T. Saleh (Ed.), *Applying Nanotechnology to the Desulfurization Process in Petroleum Engineering* (pp. 281–336). Hershey, PA: IGI Global. doi:10.4018/978-1-4666-9545-0.ch010

Ong, P., & Kohshelan, S. (2016). Performances of Adaptive Cuckoo Search Algorithm in Engineering Optimization. In P. Vasant, G. Weber, & V. Dieu (Eds.), *Handbook of Research on Modern Optimization Algorithms and Applications in Engineering and Economics* (pp. 676–699). Hershey, PA: IGI Global. doi:10.4018/978-1-4666-9644-0.ch026

Osho, M. B. (2018). Industrial Enzyme Technology: Potential Applications. In S. Bharati & P. Chaurasia (Eds.), *Research Advancements in Pharmaceutical, Nutritional, and Industrial Enzymology* (pp. 375–394). Hershey, PA: IGI Global. doi:10.4018/978-1-5225-5237-6.ch017

Padmaja, P., & Marutheswar, G. (2017). Certain Investigation on Secured Data Transmission in Wireless Sensor Networks. *International Journal of Mobile Computing and Multimedia Communications*, *8*(1), 48–61. doi:10.4018/IJMCMC.2017010104

Paixão, S. M., Silva, T. P., Arez, B. F., & Alves, L. (2016). Advances in the Reduction of the Costs Inherent to Fossil Fuels' Biodesulfurization towards Its Potential Industrial Application. In T. Saleh (Ed.), *Applying Nanotechnology to the Desulfurization Process in Petroleum Engineering* (pp. 390–425). Hershey, PA: IGI Global. doi:10.4018/978-1-4666-9545-0.ch013

Palmer, S., & Hall, W. (2017). An Evaluation of Group Work in First-Year Engineering Design Education. In R. Tucker (Ed.), *Collaboration and Student Engagement in Design Education* (pp. 145–168). Hershey, PA: IGI Global. doi:10.4018/978-1-5225-0726-0.ch007

Panneer, R. (2017). Effect of Composition of Fibers on Properties of Hybrid Composites. *International Journal of Manufacturing, Materials, and Mechanical Engineering*, *7*(4), 28–43. doi:10.4018/IJMMME.2017100103

Parker, J. (2016). Hubble's Expanding Universe: A Model for Quality in Technology Infused engineering and Technology Education. *International Journal of Quality Assurance in Engineering and Technology Education*, *5*(2), 16–29. doi:10.4018/IJQAETE.2016040102

Paul, S., & Roy, P. (2018). Optimal Design of Power System Stabilizer Using a Novel Evolutionary Algorithm. *International Journal of Energy Optimization and Engineering*, *7*(3), 24–46. doi:10.4018/IJEOE.2018070102

Pavaloiu, A. (2018). Artificial Intelligence Ethics in Biomedical-Engineering-Oriented Problems. In U. Kose, G. Guraksin, & O. Deperlioglu (Eds.), *Nature-Inspired Intelligent Techniques for Solving Biomedical Engineering Problems* (pp. 219–231). Hershey, PA: IGI Global. doi:10.4018/978-1-5225-4769-3.ch010

Peña, F. (2016). A Semi-Discrete Approach for the Numerical Simulation of Freestanding Blocks. In V. Sarhosis, K. Bagi, J. Lemos, & G. Milani (Eds.), *Computational Modeling of Masonry Structures Using the Discrete Element Method* (pp. 416–439). Hershey, PA: IGI Global. doi:10.4018/978-1-5225-0231-9.ch016

Penchovsky, R., & Traykovska, M. (2016). Synthetic Approaches to Biology: Engineering Gene Control Circuits, Synthesizing, and Editing Genomes. In M. Bououdina (Ed.), *Emerging Research on Bioinspired Materials Engineering* (pp. 323–351). Hershey, PA: IGI Global. doi:10.4018/978-1-4666-9811-6.ch012

Pieroni, A., & Iazeolla, G. (2016). Engineering QoS and Energy Saving in the Delivery of ICT Services. In P. Vasant & N. Voropai (Eds.), *Sustaining Power Resources through Energy Optimization and Engineering* (pp. 208–226). Hershey, PA: IGI Global. doi:10.4018/978-1-4666-9755-3.ch009

Pioro, I., Mahdi, M., & Popov, R. (2017). Application of Supercritical Pressures in Power Engineering. In L. Chen & Y. Iwamoto (Eds.), *Advanced Applications of Supercritical Fluids in Energy Systems* (pp. 404–457). Hershey, PA: IGI Global. doi:10.4018/978-1-5225-2047-4.ch013

Plaksina, T., & Gildin, E. (2017). Rigorous Integrated Evolutionary Workflow for Optimal Exploitation of Unconventional Gas Assets. *International Journal of Energy Optimization and Engineering*, *6*(1), 101–122. doi:10.4018/IJEOE.2017010106

Puppala, A. J., Bheemasetti, T. V., Zou, H., Yu, X., Pedarla, A., & Cai, G. (2016). Spatial Variability Analysis of Soil Properties using Geostatistics. In P. Samui (Ed.), *Handbook of Research on Advanced Computational Techniques for Simulation-Based Engineering* (pp. 195–226). Hershey, PA: IGI Global. doi:10.4018/978-1-4666-9479-8.ch008

Ramdani, N., & Azibi, M. (2018). Polymer Composite Materials for Microelectronics Packaging Applications: Composites for Microelectronics Packaging. In K. Kumar & J. Davim (Eds.), *Composites and Advanced Materials for Industrial Applications* (pp. 177–211). Hershey, PA: IGI Global. doi:10.4018/978-1-5225-5216-1.ch009

Ramesh, M., Garg, R., & Subrahmanyam, G. V. (2017). Investigation of Influence of Quenching and Annealing on the Plane Fracture Toughness and Brittle to Ductile Transition Temperature of the Zinc Coated Structural Steel Materials. *International Journal of Surface Engineering and Interdisciplinary Materials Science*, *5*(2), 33–42. doi:10.4018/IJSEIMS.2017070103

Razavi, A. M., & Ahmad, R. (2016). Agile Software Development Challenges in Implementation and Adoption: Focusing on Large and Distributed Settings – Past Experiences, Emergent Topics. In I. Ghani, D. Jawawi, S. Dorairaj, & A. Sidky (Eds.), *Emerging Innovations in Agile Software Development* (pp. 175–207). Hershey, PA: IGI Global. doi:10.4018/978-1-4666-9858-1.ch010

Reccia, E., Cecchi, A., & Milani, G. (2016). FEM/DEM Approach for the Analysis of Masonry Arch Bridges. In V. Sarhosis, K. Bagi, J. Lemos, & G. Milani (Eds.), *Computational Modeling of Masonry Structures Using the Discrete Element Method* (pp. 367–392). Hershey, PA: IGI Global. doi:10.4018/978-1-5225-0231-9.ch014

Ro, H. K., & McIntosh, K. (2016). Constructing Conducive Environment for Women of Color in Engineering Undergraduate Education. In U. Thomas & J. Drake (Eds.), *Critical Research on Sexism and Racism in STEM Fields* (pp. 23–48). Hershey, PA: IGI Global. doi:10.4018/978-1-5225-0174-9.ch002

Rodulfo-Baechler, S. M. (2016). Dual Role of Perovskite Hollow Fiber Membrane in the Methane Oxidation Reactions. In H. Al-Megren & T. Xiao (Eds.), *Petrochemical Catalyst Materials, Processes, and Emerging Technologies* (pp. 385–430). Hershey, PA: IGI Global. doi:10.4018/978-1-4666-9975-5.ch014

Rudolf, S., Biryuk, V. V., & Volov, V. (2018). Vortex Effect, Vortex Power: Technology of Vortex Power Engineering. In V. Kharchenko & P. Vasant (Eds.), *Handbook of Research on Renewable Energy and Electric Resources for Sustainable Rural Development* (pp. 500–533). Hershey, PA: IGI Global. doi:10.4018/978-1-5225-3867-7.ch021

Sah, A., Bhadula, S. J., Dumka, A., & Rawat, S. (2018). A Software Engineering Perspective for Development of Enterprise Applications. In A. Elçi (Ed.), *Handbook of Research on Contemporary Perspectives on Web-Based Systems* (pp. 1–23). Hershey, PA: IGI Global. doi:10.4018/978-1-5225-5384-7.ch001

Sahoo, P., & Roy, S. (2017). Tribological Behavior of Electroless Ni-P, Ni-P-W and Ni-P-Cu Coatings: A Comparison. *International Journal of Surface Engineering and Interdisciplinary Materials Science, 5*(1), 1–15. doi:10.4018/IJSEIMS.2017010101

Sahoo, S. (2018). Laminated Composite Hypar Shells as Roofing Units: Static and Dynamic Behavior. In K. Kumar & J. Davim (Eds.), *Composites and Advanced Materials for Industrial Applications* (pp. 249–269). Hershey, PA: IGI Global. doi:10.4018/978-1-5225-5216-1.ch011

Sahu, H., & Hungyo, M. (2018). Introduction to SDN and NFV. In A. Dumka (Ed.), *Innovations in Software-Defined Networking and Network Functions Virtualization* (pp. 1–25). Hershey, PA: IGI Global. doi:10.4018/978-1-5225-3640-6.ch001

Saikia, P., Bharadwaj, S. K., & Miah, A. T. (2016). Peroxovanadates and Its Bio-Mimicking Relation with Vanadium Haloperoxidases. In M. Bououdina (Ed.), *Emerging Research on Bioinspired Materials Engineering* (pp. 197–219). Hershey, PA: IGI Global. doi:10.4018/978-1-4666-9811-6.ch007

Saladino, R., Botta, G., & Crucianelli, M. (2016). Advances in Nanotechnology Transition Metal Catalysts in Oxidative Desulfurization (ODS) Processes: Nanotechnology Applied to ODS Processing. In T. Saleh (Ed.), *Applying Nanotechnology to the Desulfurization Process in Petroleum Engineering* (pp. 180–215). Hershey, PA: IGI Global. doi:10.4018/978-1-4666-9545-0.ch006

Saleh, T. A., Danmaliki, G. I., & Shuaib, T. D. (2016). Nanocomposites and Hybrid Materials for Adsorptive Desulfurization. In T. Saleh (Ed.), *Applying Nanotechnology to the Desulfurization Process in Petroleum Engineering* (pp. 129–153). Hershey, PA: IGI Global. doi:10.4018/978-1-4666-9545-0.ch004

Saleh, T. A., Shuaib, T. D., Danmaliki, G. I., & Al-Daous, M. A. (2016). Carbon-Based Nanomaterials for Desulfurization: Classification, Preparation, and Evaluation. In T. Saleh (Ed.), *Applying Nanotechnology to the Desulfurization Process in Petroleum Engineering* (pp. 154–179). Hershey, PA: IGI Global. doi:10.4018/978-1-4666-9545-0.ch005

Salem, A. M., & Shmelova, T. (2018). Intelligent Expert Decision Support Systems: Methodologies, Applications, and Challenges. In T. Shmelova, Y. Sikirda, N. Rizun, A. Salem, & Y. Kovalyov (Eds.), *Socio-Technical Decision Support in Air Navigation Systems: Emerging Research and Opportunities* (pp. 215–242). Hershey, PA: IGI Global. doi:10.4018/978-1-5225-3108-1.ch007

Samal, M. (2017). FE Analysis and Experimental Investigation of Cracked and Un-Cracked Thin-Walled Tubular Components to Evaluate Mechanical and Fracture Properties. In P. Samui, S. Chakraborty, & D. Kim (Eds.), *Modeling and Simulation Techniques in Structural Engineering* (pp. 266–293). Hershey, PA: IGI Global. doi:10.4018/978-1-5225-0588-4.ch009

Samal, M., & Balakrishnan, K. (2017). Experiments on a Ring Tension Setup and FE Analysis to Evaluate Transverse Mechanical Properties of Tubular Components. In P. Samui, S. Chakraborty, & D. Kim (Eds.), *Modeling and Simulation Techniques in Structural Engineering* (pp. 91–115). Hershey, PA: IGI Global. doi:10.4018/978-1-5225-0588-4.ch004

Santhanakumar, M., Adalarasan, R., & Rajmohan, M. (2016). An Investigation in Abrasive Waterjet Cutting of Al6061/SiC/Al2O3 Composite Using Principal Component Based Response Surface Methodology. *International Journal of Manufacturing, Materials, and Mechanical Engineering*, *6*(4), 30–47. doi:10.4018/IJMMME.2016100103

Sareen, N., & Bhattacharya, S. (2016). Cleaner Energy Fuels: Hydrodesulfurization and Beyond. In T. Saleh (Ed.), *Applying Nanotechnology to the Desulfurization Process in Petroleum Engineering* (pp. 84–128). Hershey, PA: IGI Global. doi:10.4018/978-1-4666-9545-0.ch003

Sarhosis, V. (2016). Micro-Modeling Options for Masonry. In V. Sarhosis, K. Bagi, J. Lemos, & G. Milani (Eds.), *Computational Modeling of Masonry Structures Using the Discrete Element Method* (pp. 28–60). Hershey, PA: IGI Global. doi:10.4018/978-1-5225-0231-9.ch002

Sarhosis, V., Oliveira, D. V., & Lourenco, P. B. (2016). On the Mechanical Behavior of Masonry. In V. Sarhosis, K. Bagi, J. Lemos, & G. Milani (Eds.), *Computational Modeling of Masonry Structures Using the Discrete Element Method* (pp. 1–27). Hershey, PA: IGI Global. doi:10.4018/978-1-5225-0231-9.ch001

Satyam, N. (2016). Liquefaction Modelling of Granular Soils using Discrete Element Method. In P. Samui (Ed.), *Handbook of Research on Advanced Computational Techniques for Simulation-Based Engineering* (pp. 381–441). Hershey, PA: IGI Global. doi:10.4018/978-1-4666-9479-8.ch015

Sawant, S. (2018). Deep Learning and Biomedical Engineering. In U. Kose, G. Guraksin, & O. Deperlioglu (Eds.), *Nature-Inspired Intelligent Techniques for Solving Biomedical Engineering Problems* (pp. 283–296). Hershey, PA: IGI Global. doi:10.4018/978-1-5225-4769-3.ch014

Sezgin, H., & Berkalp, O. B. (2018). Textile-Reinforced Composites for the Automotive Industry. In K. Kumar & J. Davim (Eds.), *Composites and Advanced Materials for Industrial Applications* (pp. 129–156). Hershey, PA: IGI Global. doi:10.4018/978-1-5225-5216-1.ch007

Shah, M. Z., Gazder, U., Bhatti, M. S., & Hussain, M. (2018). Comparative Performance Evaluation of Effects of Modifier in Asphaltic Concrete Mix. *International Journal of Strategic Engineering*, *1*(2), 13–25. doi:10.4018/IJoSE.2018070102

Shah, V. S., Shah, H. R., & Samui, P. (2016). Application of Meta-Models (MPMR and ELM) for Determining OMC, MDD and Soaked CBR Value of Soil. In S. Bhattacharyya, P. Banerjee, D. Majumdar, & P. Dutta (Eds.), *Handbook of Research on Advanced Hybrid Intelligent Techniques and Applications* (pp. 454–482). Hershey, PA: IGI Global. doi:10.4018/978-1-4666-9474-3.ch015

Sharma, N., & Kumar, K. (2018). Fabrication of Porous NiTi Alloy Using Organic Binders. In K. Kumar & J. Davim (Eds.), *Composites and Advanced Materials for Industrial Applications* (pp. 38–62). Hershey, PA: IGI Global. doi:10.4018/978-1-5225-5216-1.ch003

Sharma, T. K. (2016). Application of Shuffled Frog Leaping Algorithm in Software Project Scheduling. In P. Saxena, D. Singh, & M. Pant (Eds.), *Problem Solving and Uncertainty Modeling through Optimization and Soft Computing Applications* (pp. 225–238). Hershey, PA: IGI Global. doi:10.4018/978-1-4666-9885-7.ch011

Shivach, P., Nautiyal, L., & Ram, M. (2018). Applying Multi-Objective Optimization Algorithms to Mechanical Engineering. In M. Ram & J. Davim (Eds.), *Soft Computing Techniques and Applications in Mechanical Engineering* (pp. 287–301). Hershey, PA: IGI Global. doi:10.4018/978-1-5225-3035-0.ch014

Shmelova, T. (2018). Stochastic Methods for Estimation and Problem Solving in Engineering: Stochastic Methods of Decision Making in Aviation. In S. Kadry (Ed.), *Stochastic Methods for Estimation and Problem Solving in Engineering* (pp. 139–160). Hershey, PA: IGI Global. doi:10.4018/978-1-5225-5045-7.ch006

Shukla, R., Anapagaddi, R., Singh, A. K., Allen, J. K., Panchal, J. H., & Mistree, F. (2016). Integrated Computational Materials Engineering for Determining the Set Points of Unit Operations for Production of a Steel Product Mix. In S. Datta & J. Davim (Eds.), *Computational Approaches to Materials Design: Theoretical and Practical Aspects* (pp. 163–191). Hershey, PA: IGI Global. doi:10.4018/978-1-5225-0290-6.ch006

Siero González, L. R., & Romo Vázquez, A. (2017). Didactic Sequences Teaching Mathematics for Engineers With Focus on Differential Equations. In M. Ramírez-Montoya (Ed.), *Handbook of Research on Driving STEM Learning With Educational Technologies* (pp. 129–151). Hershey, PA: IGI Global. doi:10.4018/978-1-5225-2026-9.ch007

Singh, R., & Dutta, S. (2018). Visible Light Active Nanocomposites for Photocatalytic Applications. In K. Kumar & J. Davim (Eds.), *Composites and Advanced Materials for Industrial Applications* (pp. 270–296). Hershey, PA: IGI Global. doi:10.4018/978-1-5225-5216-1.ch012

Singh, R., & Lou, H. H. (2016). Safety and Efficiency Enhancement in LNG Terminals. In H. Al-Megren & T. Xiao (Eds.), *Petrochemical Catalyst Materials, Processes, and Emerging Technologies* (pp. 164–176). Hershey, PA: IGI Global. doi:10.4018/978-1-4666-9975-5.ch007

Sözbilir, H., Özkaymak, Ç., Uzel, B., & Sümer, Ö. (2018). Criteria for Surface Rupture Microzonation of Active Faults for Earthquake Hazards in Urban Areas. In N. Ceryan (Ed.), *Handbook of Research on Trends and Digital Advances in Engineering Geology* (pp. 187–230). Hershey, PA: IGI Global. doi:10.4018/978-1-5225-2709-1.ch005

Stanciu, I. (2018). Stochastic Methods in Microsystems Engineering. In S. Kadry (Ed.), *Stochastic Methods for Estimation and Problem Solving in Engineering* (pp. 161–176). Hershey, PA: IGI Global. doi:10.4018/978-1-5225-5045-7.ch007

Strebkov, D., Nekrasov, A., Trubnikov, V., & Nekrasov, A. (2018). Single-Wire Resonant Electric Power Systems for Renewable-Based Electric Grid. In V. Kharchenko & P. Vasant (Eds.), *Handbook of Research on Renewable Energy and Electric Resources for Sustainable Rural Development* (pp. 449–474). Hershey, PA: IGI Global. doi:10.4018/978-1-5225-3867-7.ch019

Subburaman, D., Jagan, J., Dalkiliç, Y., & Samui, P. (2016). Reliability Analysis of Slope Using MPMR, GRNN and GPR. In F. Miranda & C. Abreu (Eds.), *Handbook of Research on Computational Simulation and Modeling in Engineering* (pp. 208–224). Hershey, PA: IGI Global. doi:10.4018/978-1-4666-8823-0.ch007

Sun, J., Wan, S., Lin, J., & Wang, Y. (2016). Advances in Catalytic Conversion of Syngas to Ethanol and Higher Alcohols. In H. Al-Megren & T. Xiao (Eds.), *Petrochemical Catalyst Materials, Processes, and Emerging Technologies* (pp. 177–215). Hershey, PA: IGI Global. doi:10.4018/978-1-4666-9975-5.ch008

Tüdeş, Ş., Kumlu, K. B., & Ceryan, S. (2018). Integration Between Urban Planning and Natural Hazards For Resilient City. In N. Ceryan (Ed.), *Handbook of Research on Trends and Digital Advances in Engineering Geology* (pp. 591–630). Hershey, PA: IGI Global. doi:10.4018/978-1-5225-2709-1.ch017

Tyukhov, I., Rezk, H., & Vasant, P. (2016). Modern Optimization Algorithms and Applications in Solar Photovoltaic Engineering. In P. Vasant & N. Voropai (Eds.), *Sustaining Power Resources through Energy Optimization and Engineering* (pp. 390–445). Hershey, PA: IGI Global. doi:10.4018/978-1-4666-9755-3.ch016

Ulamis, K. (2018). Soil Liquefaction Assessment by Anisotropic Cyclic Triaxial Test. In N. Ceryan (Ed.), *Handbook of Research on Trends and Digital Advances in Engineering Geology* (pp. 631–664). Hershey, PA: IGI Global. doi:10.4018/978-1-5225-2709-1.ch018

Umar, M. A., Tenuche, S. S., Yusuf, S. A., Abdulsalami, A. O., & Kufena, A. M. (2016). Usability Engineering in Agile Software Development Processes. In I. Ghani, D. Jawawi, S. Dorairaj, & A. Sidky (Eds.), *Emerging Innovations in Agile Software Development* (pp. 208–221). Hershey, PA: IGI Global. doi:10.4018/978-1-4666-9858-1.ch011

Üzüm, O., & Çakır, Ö. A. (2016). A Bio-Inspired Phenomena in Cementitious Materials: Self-Healing. In M. Bououdina (Ed.), *Emerging Research on Bioinspired Materials Engineering* (pp. 220–246). Hershey, PA: IGI Global. doi:10.4018/978-1-4666-9811-6.ch008

Valente, M., & Milani, G. (2017). Seismic Assessment and Retrofitting of an Under-Designed RC Frame Through a Displacement-Based Approach. In V. Plevris, G. Kremmyda, & Y. Fahjan (Eds.), *Performance-Based Seismic Design of Concrete Structures and Infrastructures* (pp. 36–58). Hershey, PA: IGI Global. doi:10.4018/978-1-5225-2089-4.ch002

Vasant, P. (2018). A General Medical Diagnosis System Formed by Artificial Neural Networks and Swarm Intelligence Techniques. In U. Kose, G. Guraksin, & O. Deperlioglu (Eds.), *Nature-Inspired Intelligent Techniques for Solving Biomedical Engineering Problems* (pp. 130–145). Hershey, PA: IGI Global. doi:10.4018/978-1-5225-4769-3.ch006

Vergara, D., Lorenzo, M., & Rubio, M. (2016). On the Use of Virtual Environments in Engineering Education. *International Journal of Quality Assurance in Engineering and Technology Education*, *5*(2), 30–41. doi:10.4018/IJQAETE.2016040103

Verrollot, J., Tolonen, A., Harkonen, J., & Haapasalo, H. J. (2018). Challenges and Enablers for Rapid Product Development. *International Journal of Applied Industrial Engineering*, *5*(1), 25–49. doi:10.4018/IJAIE.2018010102

Wagner, C., & Ryan, C. (2016). Physical and Digital Integration Strategies of Electronic Device Supply Chains and Their Applicability to ETO Supply Chains. In R. Addo-Tenkorang, J. Kantola, P. Helo, & A. Shamsuzzoha (Eds.), *Supply Chain Strategies and the Engineer-to-Order Approach* (pp. 224–245). Hershey, PA: IGI Global. doi:10.4018/978-1-5225-0021-6.ch011

Wang, Z., Wu, P., Lan, L., & Ji, S. (2016). Preparation, Characterization and Desulfurization of the Supported Nickel Phosphide Catalysts. In H. Al-Megren & T. Xiao (Eds.), *Petrochemical Catalyst Materials, Processes, and Emerging Technologies* (pp. 431–458). Hershey, PA: IGI Global. doi:10.4018/978-1-4666-9975-5.ch015

Yardimci, A. G., & Karpuz, C. (2018). Fuzzy Rock Mass Rating: Soft-Computing-Aided Preliminary Stability Analysis of Weak Rock Slopes. In N. Ceryan (Ed.), *Handbook of Research on Trends and Digital Advances in Engineering Geology* (pp. 97–131). Hershey, PA: IGI Global. doi:10.4018/978-1-5225-2709-1.ch003

Zhang, L., Ding, S., Sun, S., Han, B., Yu, X., & Ou, J. (2016). Nano-Scale Behavior and Nano-Modification of Cement and Concrete Materials. In A. Khitab & W. Anwar (Eds.), *Advanced Research on Nanotechnology for Civil Engineering Applications* (pp. 28–79). Hershey, PA: IGI Global. doi:10.4018/978-1-5225-0344-6.ch002

Zindani, D., & Kumar, K. (2018). Industrial Applications of Polymer Composite Materials. In K. Kumar & J. Davim (Eds.), *Composites and Advanced Materials for Industrial Applications* (pp. 1–15). Hershey, PA: IGI Global. doi:10.4018/978-1-5225-5216-1.ch001

Zindani, D., Maity, S. R., & Bhowmik, S. (2018). A Decision-Making Approach for Material Selection of Polymeric Composite Bumper Beam. In K. Kumar & J. Davim (Eds.), *Composites and Advanced Materials for Industrial Applications* (pp. 112–128). Hershey, PA: IGI Global. doi:10.4018/978-1-5225-5216-1.ch006

About the Contributors

Soubantika Palchoudhury is an Assistant Professor in the Civil and Chemical Engineering Department of the University of Tennessee at Chattanooga. Her research focuses on developing new nanoscale architectures, understanding their structure-property relations, and their interactions with the environment. She has been a faculty member at UTC since 2015. Prior to joining UTC, she did three years of postdoctoral research at The University of Alabama, the University of South Carolina, and at Yale University. She earned her Ph.D. in Chemical Engineering from The University of Alabama in 2012 and bachelors in Chemical Engineering from National Institute of Technology, Durgapur, India in 2008. Her work aims to integrate material science and biological methods to engineer smart nanoparticles for therapeutic, energy, and environmental applications. She has published 26 papers and four book chapters. Her research has featured as a cover art in *Chem Comm* and has been highlighted in *Nature*. She has won state grants and has several established national and international research collaborations. She is a reviewer for a number of nanoscience journals such as *APL, JAP, Dalton Transac., Langmuir, RSC Adv.,* and *ES&T*.

* * *

Sami D. Alaruri holds a B.Sc. degree in Physics with minor in Mathematics from Kuwait University, an M.Sc. in Lasers and their Applications from University of Essex, England, an M.S., in Applied Optics and an M.S. in Engineering Management from Rose-Hulman Institute of Technology, and an M.S. in Analytical Chemistry from Purdue University at Indianapolis. Sami worked for Beckman Coulter, Inc. (Danaher Corp.), Corning, Inc., Rolls-Royce Corporation, Vaser, Inc., and Kuwait Institute for Scientific Research. Sami's main research interests are in areas related to the design and development of lasers, laser amplifiers and optical instruments. He has four issued patents. http://spie.org/profile/Sami.Alaruri-150970; https://scholar.google.com/citations?user=CwvUlGQAAAAJ&hl=en.

Renato Rubim Ribeiro de Almeida obtained the Chemistry degree in 2002 from the Federal Fluminense University (Brazil) and the master's degree in the Post-Graduate Course in Metrology for Industrial Quality from Inmetro in 2016. Experience in Chemical Metrology with emphasis on environmental and food analysis.

Cristiane Rodrigues Augusto obtained the Bachelor of Chemistry Industrial degree in 2002 from the Federal University of Rio de Janeiro (Brazil), the Master degree in 2007 and the Doctor degree in 2011, in the Technology of Chemistry and Biochemistry Process, from the Federal University of Rio de Janeiro (Brazil). Currently, the researcher works in the field of gas analysis and is the head of the Gas Analysis Laboratory (Lanag) of the Chemistry and Thermal Metrology Division (Dimqt).

Janaina Marques Rodrigues Caixeiro obtained the Bachelor of Chemistry Engineer in 1997, M.Sc. and D.Sc. in Organic Chemistry from the Federal Rural University of Rio de Janeiro, Brazil. Presently, she has been working as a researcher at National Institute of Metrology, Quality and Technology (INMETRO). The main research areas of Dra Rodrigues are focused on the development and validation of analytical methods for analysis or organic and inorganic compounds in soils, food, biological and environmental matrices.

Lucas Junqueira de Carvalho obtained the Chemistry degree in 2010 from the Fluminense Federal University of Niteroi – Rio de Janeiro (Brazil), the Master degree in 2015, in the Metrology and Analytical Chemistry area, from the Inmetro (Brazil). He works in the organic analysis laboratory (Labor) of the Chemical and Thermal Metrology Divison (Dimqt). Currently, the researcher works on the development of organic certified reference materials (CRM) in food and environmental.

Valnei Smarçaro da Cunha obtained his bachelor degree in chemical engineering from the Federal Rural University of Rio de Janeiro in 1995. Had his master's degree and doctorate from PEQ/COPPE/Federal University of Rio de Janeiro (1997 and 2001) in the area of separation processes and polymers. Has experience in chemical engineering, with emphasis on non-conventional separation processes, analytical chemistry, certified reference materials, metrology and quality. Is currently a researcher at the National Institute of Metrology, Quality and Technology-INMETRO and chief of the Chemical and Thermal Metrology Division.

Luiz Antônio d`Avila holds a degree in Chemistry (1970) and a doctorate in Organic Chemistry (1985) from the Institute of Chemistry at the Federal University of Rio de Janeiro (IQ/UFRJ). Specializations at: Laboratoire de Chimie Analytique Physique at Ecole Polytechnique – France (1980 – 1982); Expochim en Genie Chimique at Ecole Nationale Superieure des Industries Chimiques (1987) and in Industrial Preparatory Chromatography at Prochrom, France (1987 and 1993). Professor at the Institute of Chemistry (IQ/UFRJ) since 1972 and at the School of Chemistry (EQ/UFRJ) since 1999, where is Full Professor, Vice Director (2002-2006) and Director (2006-2009). General coordinator of LABCOM since 1999, the Laboratory for Fuels and Oil Byproducts, accredited by NBR ISO/IEC 17025 since 2005.

Lowell Diworth is a lecturer and consultant in the department of Pathology at the University of the West Indies, Mona Campus. His interests are, carbohydrate and lipid metabolism, diabetes mellitus, food biochemistry, as well as studying biomarkers of kidney disease.

Marcelo de Almeida Dominguez obtained the bachelor of Industrial Chemistry from Federal University Fluminense (1998), Masters in Geochemistry from Federal University Fluminense (2001) and Ph.D. in Environmental Geochemistry from Federal University Fluminense (2005). Currently, the researcher works in the Laboratory of Inorganic Analysis (Labin) Division of Chemical Metrology (Dquim) of National Institute of Metrology, Quality and Technology - INMETRO, on the development and maintenance of certified reference materials. Has experience in the area of chemistry, with a focus on Trace Analysis, Environmental Chemistry and Chemical Metrology.

Andreia de Lima Fioravante obtained the Bachelor of Chemistry Engineer degree in 2003 from the Federal University of Rio de Janeiro (Brazil) and the Master degree in 2006, in the Technology of Chemistry and Biochemistry Process, from the Federal University of Rio de Janeiro (Brazil). Currently, the researcher works in the Gas Analysis Laboratory (Lanag) of the Chemistry and Thermal Metrology Division (Dimqt) on the development and maintenance of gaseous certified reference materials, ranging from environmental gases to natural gas and ethanol (evidential purposes) primary mixtures. Also works in the technical coordination for environmental area developments of the NMI Chemistry Division.

Natwaine Sherune Gardner received her PhD in Biotechnology (high commendation) from the University of the West Indies, Mona in 2014. She is currently a lecturer in Biochemistry at the University of Technology where she sits on numerous committees. She consults in areas relating to Substances of Abuse and Human Health where her passion lies.

Fabiano Barbieri Gonzaga obtained the Bachelor of Chemistry degree in 2000 from the Federal University of Viçosa (Brazil) and the D.Sc. degree in 2006, in the Analytical Chemistry area, from the State University of Campinas (Brazil). Currently, he works in the Electrochemical Metrology area.

S. B. Jaju is working as Professor and Dean R&D at G.H. Raisoni College of Engineering, Nagpur, India. He has teaching experience of 16 years in the field of mechanical engineering and a number of publications in international journals as well as conferences.

Vinoth Kumar K. was born on 30th April 1984, in Vellore, Tamil Nadu, India. He received his Undergraduate degree in Electrical and Electronics Engineering from Anna University, Chennai, India in the year 2006 and Post Graduate degree specializing in Power Electronics and Drives from VIT University, Vellore, India in the year 2008. He is currently Assistant Professor in the Department of Electrical Sciences at Karunya University, Coimbatore, India. His research interests include Fault diagnosis in industrial drives, Neuro Fuzzy computing and Power Electronics & Drives. He has been actively involved in the Internal Quality Assurance Cell at Karunya University. His contribution in setting up the experiments for the Power Electronics Laboratory is also commendable. He published research papers in both international and national journals as well as in conferences. He is authored the Book titled "Basic Electrical and Electronics Engineering" published by John Wiley & Sons India in 2014 and Neural Networks, Soft Computing textbooks in 2012. He has delivered lectures on Energy Conservation in Industries, New techniques for Energy auditing and other related areas in various workshop and seminars as well as guided 14 UG and 14 PG Student projects and One International student project.

Mithilesh Kumar is currently working as a Principle at GEC Jhalawar Rajasthan. He has more than 20 years academic and research experience and completed M.Tech. & Ph.D. in Microwaves and antenna design field from IIT, Delhi. His research area is Microwaves and Antenna Design. He has also published more than 50 research papers in reputed national and international journals.

R. R. Lakhe is the Director of Shreyas Quality Management System, Nagpur, India. He is running his own firm providing the consultancy services in ISO, TQM, NABL accreditation, quality audit, and TPM. He has completed lots of government and private sectors projects successfully on his name and having vast experience of 25 years in the field of quality engineering.

Henry Lowe is a scientist who specializes in Medicinal chemistry, has contributed approximately 50 years in the field of science, technology, energy, the environment, wellness and health sciences nationally, regionally and internationally.

Evelyn de Freitas Magalhães obtained the Chemistry degree in 2003 from the Federal University of Rio de Janeiro (Brazil), the Master degree in 2005 and the Doctorate Degree in 2013, in the Analytical Chemistry area, from the Military Institute of Engineering (Brazil). She works in the organic analysis laboratory (Labor) of the chemistry Metrology Division (Dquim). Currently, the researcher works on the development of organic certified reference materials (CRM) in food and environmental.

Nitin K. Mandavgade is a faculty of RTMNU, Nagpur University, Nagpur having experience of 18 years. He had total 48 publications in various reputed international journals as well as conferences. His area of research is uncertainty of measurement.

Alessandro Massaro received the Laurea degree in electronic engineering and the Ph.D. degree in telecommunication engineering from the Università Politecnica delle Marche, Ancona, Italy, in 2001 and 2004, respectively. From 2004 to 2006 he worked as Research Scientist (post-doc) in the Department of Electromagnetism and Bioengineering at Università Politecnica delle Marche. In 2006, he worked in R&D in a medical industry (endoscope design and optical systems). He worked for two years with National Nanotechnology Laboratory of CNR-INFM, Università del Salento, as Principal Investigator. He was team leader in Robotics Lab. 1 and Smart Material Lab of the Center of Bio-Molecular Nanotechnology of Italian Institute Technology (IIT), Arnesano, Lecce, Italy. He was CNR researcher in Aerospace applications (CNR- NANOTEC). Currently, he is R&D chief in Dyrecta Lab s.r.l.

Donovan A. McGrowder is a Senior Lecturer and Head of the Sub-Department of Chemical Pathology, Department of Pathology, Faculty of Medical Sciences, The University of the West Indies, Mona Campus.

Laura Alves das Neves obtained the Chemistry degree in 2001 from the Federal University of Rio de Janeiro (Brazil) and the Master degree in 2005 from Military Institute of Engineering (Brazil). Experience in Chemical Metrology with emphasis on the following topics: traceability, primary methods of measurement, certified reference materials, validation methodologies and analysis of environmental contaminants, proficiency testing and international comparisons.

Eliane Cristina Pires do Rego obtained the Chemistry degree in 2003 from the State University of Rio de Janeiro (Brazil), the Master degree in 2007 and the Doctorate Degree in 2015, in the Analytical Chemistry area, from the Federal Fluminense University (Brazil). Currently, the researcher is the head of the Organic Analysis Laboratory (Labor) of the Chemical and Thermal Metrology Divison (Dimqt) and works on the development of organic certified reference materials (CRM) in food, environmental, biofuel, pharmaceutical, forensic and health areas. Also works in Proficiency Testing scheme (PTS) and method validation training.

Claudia Cipriano Ribeiro obtained the Technologist of Environment degree in 2002 from the Federal Center of Technological Education CelsoSuckow da Fonseca (Brazil) and the Specialist degree in 2004, in Sanitary Engineering and Environment Control, from the National School of Public Health SérgioArouca (Brazil). Currently, the researcher works in the Gas Analysis Laboratory (Lanag) of the Chemistry Metrology Division (Dimqt) on the development and maintenance of gaseous certified reference materials, as well as performing proficiency tests for the gas area, ranging from environmental gases to natural gas and automotive emissions primary mixtures.

Elaine Batista de Santana obtained the technical in metrology degree in 2010 from the Inmetro/SEE-RJ/CECO technical course, the Chemistry degree in 2011, from the Grande Rio University (Brazil) and the master's degree in the Post-Graduate Course in Metrology for Industrial Quality from Inmetro. Currently, she works in the organic analysis laboratory (Labor) of the Chemical and Thermal Metrology Division (Dimqt).

Rodrigo C de Sena received his B.Sc. in Chemistry, M.Sc. in organic chemistry and D.Sc. in chemistry from the Federal University of Rio de Janeiro, Brazil. Presently, he has been working as a researcher at National Institute of Metrology, Quality, and Technology (INMETRO). The main research areas of Dr. de Sena are focused on the development of analytical methods for analysis of toxic and essential elements in soil, food, fuel, biological and environmental matrices.

Satya P. Singh is currently working as a postdoctoral fellow at the Nanyang Technological University, Singapore in the area of image processing.

Sidney P. Sobral obtained the Bachelor of Chemistry degree in 2004 from the Souza Marques Technical-Educational Foundation (Brazil). Currently, he works in the Electrochemical Metrology area.

Index

N

P

R

T

U

V

W

www.ingramcontent.com/pod-product-compliance
Lightning Source LLC
LaVergne TN
LVHW061535290926
849162LV00004B/799